Diverse Applications of Organic-Inorganic Nanocomposites:

Emerging Research and Opportunities

Gabriele Clarizia
Institute on Membrane Technology, National Research Council, Italy

Paola Bernardo
Institute on Membrane Technology, National Research Council, Italy

A volume in the Advances in Mechatronics and Mechanical Engineering (AMME) Book Series

Published in the United States of America by
IGI Global
Engineering Science Reference (an imprint of IGI Global)
701 E. Chocolate Avenue
Hershey PA, USA 17033
Tel: 717-533-8845
Fax: 717-533-8661
E-mail: cust@igi-global.com
Web site: http://www.igi-global.com

Library of Congress Cataloging-in-Publication Data

Names: Clarizia, Gabriele, 1970- editor. | Bernardo, Paola, 1973- editor.
Title: Diverse applications of organic-inorganic nanocomposites : emerging research and opportunities / Gabriele Clarizia and Paola Bernardo, editors.
Description: Hershey, PA : Engineering Science Reference, an imprint of IGI Global, 2019. | Includes bibliographical references and index. | Summary: "This book explores the application of organic-inorganic nanocomposites in human activities"-- Provided by publisher.
Identifiers: LCCN 2019030380 (print) | LCCN 2019030381 (ebook) | ISBN 9781799815303 (hardcover) | ISBN 9781799815310 (paperback) | ISBN 9781799815327 (ebook)
Subjects: LCSH: Nanocomposites (Materials)--Industrial applications.
Classification: LCC TA418.9.N35 D58 2019 (print) | LCC TA418.9.N35 (ebook) | DDC 620.1/18--dc23
LC record available at https://lccn.loc.gov/2019030380
LC ebook record available at https://lccn.loc.gov/2019030381

This book is published in the IGI Global book series Advances in Mechatronics and Mechanical Engineering (AMME) (ISSN: 2328-8205; eISSN: 2328-823X)

British Cataloguing in Publication Data
A Cataloguing in Publication record for this book is available from the British Library.

Advances in Mechatronics and Mechanical Engineering (AMME) Book Series

ISSN:2328-8205
EISSN:2328-823X

Editor-in-Chief: J. Paulo Davim University of Aveiro, Portugal

MISSION

With its aid in the creation of smartphones, cars, medical imaging devices, and manufacturing tools, the mechatronics engineering field is in high demand. Mechatronics aims to combine the principles of mechanical, computer, and electrical engineering together to bridge the gap of communication between the different disciplines.

The **Advances in Mechatronics and Mechanical Engineering (AMME) Book Series** provides innovative research and practical developments in the field of mechatronics and mechanical engineering. This series covers a wide variety of application areas in electrical engineering, mechanical engineering, computer and software engineering; essential for academics, practitioners, researchers, and industry leaders.

COVERAGE

- Sustainable and green manufacturing
- Control Methodologies
- Bioengineering Materials
- Vibration and acoustics
- Intelligent Navigation
- Control Systems Modelling and Analysis
- Tribology and surface engineering
- Intelligent Sensing
- Nanomaterials and nanomanufacturing
- Design and Manufacture

IGI Global is currently accepting manuscripts for publication within this series. To submit a proposal for a volume in this series, please contact our Acquisition Editors at Acquisitions@igi-global.com or visit: http://www.igi-global.com/publish/.

The Advances in Mechatronics and Mechanical Engineering (AMME) Book Series (ISSN 2328-8205) is published by IGI Global, 701 E. Chocolate Avenue, Hershey, PA 17033-1240, USA, www.igi-global.com. This series is composed of titles available for purchase individually; each title is edited to be contextually exclusive from any other title within the series. For pricing and ordering information please visit http://www.igi-global.com/book-series/advances-mechatronics-mechanical-engineering/73808. Postmaster: Send all address changes to above address.

Titles in this Series

For a list of additional titles in this series, please visit:
https://www.igi-global.com/book-series/advances-mechatronics-mechanical-engineering/73808

Handbook of Research on Artificial Intelligence Applications in the Aviation and Aerospace Indstries
Tetiana Shmelova (National Aviation University, Ukraine) Yuliya Sikirda (Flight Academy of National Aviation University, Ukraine) and Arnold Sterenharz (EXOLAUNCH GmbH, Germany)
Engineering Science Reference • © 2020 • 517pp • H/C (ISBN: 9781799814153) • US $295.00

Design and Optimization of Sensors and Antennas for Wearable Devices Emerging Research and Opportunities
Vinod Kumar Singh (S. R. Group of Institutions Jhansi, India) Ratnesh Tiwari (Bhilai Institute of Technology, India) Vikas Dubey (Bhilai Institute of Technology, India) Zakir Ali (IET Bundelkhand University, India) and Ashutosh Kumar Singh (Indian Institute of Information Technology, ndia)
Engineering Science Reference • © 2020 • 196pp • H/C (ISBN: 9781522596837) • US $215.00

Global Advancements in Connected and Intelligent Mobility Emerging Research and Opportunities
Fatma Outay (Zayed University, UAE) Ansar-Ul-Haque Yasar (Hasselt University, Belgium) and Elhadi Shakshuki (Acadia University, Canada)
Engineering Science Reference • © 2020 • 278pp • H/C (ISBN: 9781522590194) • US $195.00

Automated Systems in the Aviation and Aerospace Industries
Tetiana Shmelova (National Aviation University, Ukraine) Yuliya Sikirda (Kirovograd Flight Academy of the National Aviation University, Ukraine) Nina Rizun (Gdansk University of Technology, Poland) Dmytro Kucherov (National Aviation University, Ukraine) and Konstantin Dergachov (National Aerospace University – Kharkiv Aviation Institute, Ukraine)

For an entire list of titles in this series, please visit:
https://www.igi-global.com/book-series/advances-mechatronics-mechanical-engineering/73808

IGI Global
DISSEMINATOR OF KNOWLEDGE

701 East Chocolate Avenue, Hershey, PA 17033, USA
Tel: 717-533-8845 x100 • Fax: 717-533-8661
E-Mail: cust@igi-global.com • www.igi-global.com

Editorial Advisory Board

Table of Contents

Preface

In a global context of sustainable development in which an always growing population has fewer resources available, the investigation of new concepts and production systems has become a primary requirement. In this particular scenario, where a greater sensitivity towards the environment and the awareness of the huge value of the natural capital must be essential cornerstones, a more rational use of natural resources is needed.

It must be said that nowadays the scientific and technological progress has reached levels once unimaginable, mainly thanks to the development of new materials capable of keeping their remarkable properties also operating in extreme conditions. The development of increasingly advanced production techniques has made possible to use these materials in increasingly thin and miniaturized forms, thus meeting the requirements of a large number of applications ranging from electronics to medicine, from the transportation to energy, passing for health and wellness. However, in many fields it seems that the physically conceivable limits have been already achieved in virtue of arrangements that cannot be further reduced in their macroscopic dimensions.

The discovery and the opportunity to take advantage of exceptional properties that lead from the synergy between heterogeneous materials, established at nanoscale level, represent the most advanced frontier in the development of nanocomposite systems with superior mechanical, thermal, electrical and transport properties.

Nanocomposites are multiphase materials available in nature or chemically synthesized, where at least a component is available at nanoscale size, independently on its shape, structure and geometry. In these multiphase materials, the interactions between the different components are enhanced by their intimate contact. Shape, aspect factor, loading, morphology are some of the distinctive features of nanofillers that are embedded in the matrix, itself already suitable to provide specific strength properties. This combination guarantees excellent macroscopic properties by taking advantage of the peculiar characteristics that nanometer-sized fillers give to a suitable matrix compared to those obtainable with the same fillers at micrometric dimensions.

Nanocomposites usually comprise strengthening layers of various materials such as clay, plastics, glass, carbon nanotubes, graphene and others. The binder material is usually a polymeric resin (e.g., epoxy) and this matrix holds the materials together to impart strength and higher toughness once the composite system has cured.

The scientific literature proposes examples of applications that cover practically all fields of technology and in view of continuous development, it is necessary to take stock of the situation in the sectors of greatest interest to better address the new discoveries and future innovations. In this perspective, the knowledge of the current market and of the distinct growth prospects in the different segments is fundamental. Packaging, aviation and automotive are some of the sectors with higher growth of nanocomposites.

Likewise, although many potential combinations may seem to be just laboratory exercises, unthinkable in real systems of great impact, understanding the principles of particle/matrix interactions is fundamental for the most appropriate choice in the specific ambit of interest.

It should be noted that the amazing properties of highly innovative materials such as graphene could not be exploited as they are in many applications, but only if dispersed in a suitable compatible matrix.

Thus, the need to achieve higher lightness and less friction in space, air and land transport applications in order to consume less fuel without losing the crucial mechanical and thermal resistances for these specific applications, can be satisfied precisely by nanocomposites.

Considering that a significant part of the food produced globally is wasted, the packaging plays a relevant role in the framework of a rational use of the resources. Food packaging requires a package that hinders the gain or loss of moisture, prevents microbial contamination, and acts as a barrier against permeation of water vapor, oxygen, carbon dioxide and other volatile compounds (e.g. flavors and taints) in addition to mechanical, optical and thermal properties. Environmentally friendly materials for food applications and to prolong the shelf life of food are desirable. Controlling the transport rate of certain species by delaying respiratory processes even with the use of substances capable of playing an antimicrobial role and, at the same time, changing the barrier properties of the conventional packaging in active packaging are the main requirements. For the final consumers of food packaged, also using nanocomposite materials, the major concern is to verify the extent of migration of nanoparticles from the package into the food and then if this migration happens, the effect of the ingestion of these nanoparticles inside the body if the nanoparticles are absorbed by the different organs and how the body metabolizes them and how and in which way the body eliminate them. However, the presence of polymer nanoparticles could significantly slow down the rate of migration of some potential hazardous chemicals, already present in the packaging films, into the food. Thus, an optimal fixation of nanoparticles in nanocomposites, including persistent suppression of oxidative damages to polymer by nanoparticles, changes of nanoparticle surface, structure or composition are mandatory requirements to be met.

Adhesion problems and compatibility of the fillers with the host matrices are two main issues to be addressed in order to guarantee a perfect dispersion of the nanofiller without defect formation at the interphase between the heterogeneous phases (nanofiller and host matrix) and channeling phenomena that lower the selectivity of the nanocomposite. The use of adhesion promoters and additives, or the chemical modification of the filler, can help in overcoming these limitations, without reduction of the ductility or processability of the final product.

In order to generate more environmentally friendly materials, as well as to decrease the dependence from the fossil-based resources, a number of biopolymers has been developed in the recent years. Biodegradable polymers typically less resistant than those of conventional synthesis can be widely applied thanks to the combination with appropriate fillers, also coming from renewable sources, that give them mechanical and thermal resistance. Accordingly, such bio-nanocomposites represent potentially high value materials of the future.

This book summarizes recent developments in the field of organic-inorganic nanocomposites, highlighting research trends and upcoming progresses. Fundamentals and application fields are provided combining the expertise of worldwide specialists. Nanocomposite preparation, characterization and modelling are addressed. Advantages and limitations are critically discussed with particular reference to more attractive examples of innovative nanofillers such as graphene and carbon nanotubes. A wide range of applications is covered, starting from wastewater treatment and exploring photocatalysis, food packaging, aerospace and automotive industry, energy and environmental applications.

The book wants to be a source of inspiration for a wide audience (both academic and industrial) involved in nanocomposites, providing a concise overview of the potential development fields for these innovative materials.

The first chapter aims to interpret and predict properties of different nanocomposites such as thermal conductivity, density, specific heat, and thermal diffusivity by using numerical models available in literature and adapted for these systems. The effect of the nanoparticles with specific shape and size, at certain loadings, on the macroscopic properties of the nanocomposite is discussed.

In Chapter 2, a brief but precise analysis of the mechanical properties of single- and multi-walled carbon nanotubes embedded in an organic matrix has been provided. The role of structural defects and segregation issues in polymer-carbon nanotubes nanocomposites are illustrated by scanning electron micrographs, mechanical and thermal analyses.

A concise overview of the graphene-based nanocomposite synthesis methods and their characterization is provided in Chapter 3. The main parameters that influence the dispersion in the polymer matrix are discussed on the basis of the most suitable preparation method depending on the filler content for the most advanced applications.

Chapter 4 focuses on synthesis and characterization of polymer/metal-oxide nanocomposites for their application to wastewater treatment. The comparison of different technologies such as membranes, biological and photochemical treatments, emphasizing the strengths and weaknesses of each of them, suggests an integration of these methods for obtaining a more efficient, optimized process.

The main aspects concerning packaging application of nanocomposites are described in Chapter 5. The effect on quality, healthy and enhanced shelf-life of the food are highlighted with reference to nanoparticles that improve mechanical, thermal and barrier properties for gases of conventional packaging materials. Innovative concepts of active packaging and nano-sensors represent the frontier for the future development and directions.

Chapter 6 discusses recent advancements in photocatalysis showing how the presence of nanocomposites can increase the performance of the actual materials in different fields by overcoming the main operating limits.

Main preparation methods of nanocomposites applied in space and automotive fields are discussed in Chapter 7. Advantages and drawbacks of nanocomposites with reference to nanofiller content and adopted synthesis protocol are presented.

The latest advances in aerospace applications are addressed in Chapter 8. Particular attention is paid to how nanocomposites have made possible to achieve extremely important results in terms of lightness and strength starting from well-known materials.

Definitely, nanocomposites represent the most attractive approach in the ambitious challenge towards materials and tools with ever better performance in different areas of human activity. This book illustrates the main paths and results of both academic and industrial research on this relevant subject, providing up-to-date references for further in-depth studies also from young investigators.

The collection in the same volume of distant but complementary research fields offers the opportunity to promote collaborations among experts with positive effects on the scientific community.

Paola Bernardo
Institute on Membrane Technology, National Research Council, Italy

Gabriele Clarizia
Institute on Membrane Technology, National Research Council, Italy

Acknowledgment

We would like to thank the publisher who gave us this opportunity, all the authors, the advisory board members and reviewers who, with their work, have enabled the implementation of this project. This undertaking has allowed us to strengthen our knowledge and our union, making us better aware of our limitations and our vocations.

Chapter 1
Numerical Study of Nanocomposites for Energy Applications

Siddhartha Kosti
https://orcid.org/0000-0001-9419-3023
Rajkiya Engineering College, Banda, India

ABSTRACT

Nanocomposites are defined as a combination of nanoparticles reinforced into the base material. They are of very small sizes (1nm = 10^{-9}m) and possesses higher thermal properties. They are widely utilized in different applications, like in energy, construction, biomedical, chemical, electronics, agriculture, cosmetics, etc. This chapter deals with the application of nanocomposites (SiC/Al_2O_3/B_4C/TiO_2/ZnO/SiO_2) in the field of energy applications by analyzing their properties (thermal-conductivity/density/specific-heat) using numerical models. The effect of nanoparticles reinforced wt. % concentration into a base material (Al6061/Al7075/H2O) is also analyzed. Results show that nanocomposites have higher effective thermal conductivity and are suitable for high heat-releasing energy devices. It is found that the addition of nanoparticles increases the surface area to volume ratio, which further increases the energy transfer rate. Results show that nanocomposites with lower effective density are suitable when there is a requirement of reduction in weight for the same heat release application.

DOI: 10.4018/978-1-7998-1530-3.ch001

INTRODUCTION

Nanocomposites are defined as a combination of nanoparticles reinforced into the base material. They are of very small sizes (1 nm = 10^{-9}m) and possesses high thermal properties like thermal conductivity, heat capacity, etc. A lot of research is going in the field of nanocomposites behavior analysis under different conditions or applications. They have been widely utilized in different applications, like in energy, construction, biomedical, chemical, electronics, agriculture, paints, and cosmetics, etc. (Sharma, 2018; Santosh, 2016).

Development in the technology is resulting in the miniaturization of energy devices, these miniature devices generate a large amount of heat which need to be released to the atmosphere for proper functioning, they also need light weight material. As capabilities of the conventional materials are limited, we need new materials that can fulfill these requirements. Nanocomposites are these new materials which can surely fulfil these requirements and can be very helpful in the energy applications, as they possess enhanced thermal properties compared to conventional materials (Lee et al, 2010). Nanocomposites are utilizing in solar and other energy conservation devices. Literature in the field of solar cooling states that use of nanocomposites increases the rate of solar cooling by a considerable amount (Al-Shamani et al, 2014). Nanoparticles like Cu and Al_2O_3 reinforcement into the base material enhances the effective thermal conductivity and energy transfer rate (Santra, 2008; Lai 2011). There are other numerous applications of the nanocomposites which have resulted in the start of new ventures/companies.

Methodology and Materials

Present work will deal with the application of nanocomposites like SiC (Silicon carbide), Aluminium oxide or alumina (Al_2O_3) and Boron carbide (B_4C), Titanium oxide (TiO_2), Zinc oxide (ZnO) and silicon oxide (SiO_2) in the field of energy. Properties of these different nanocomposites like thermal conductivity, density, specific heat, and thermal diffusivity are analyzed in detail with the help of numerical models available in the literature. Effect of addition of concentration of nanoparticles on the different base material like (Al6061/Al7075/H_2O) will also be analyzed. Table 1 represents the properties of these nanoparticles and base materials (Kosti, 2019).

Wide varieties of numerical models are available in the literature (Lee et al, 2010) to calculate the effective properties of the nanocomposites like thermal conductivity, specific heat, and density, but most of the studied area devoted towards the finding of effective thermal conductivity, as thermal conductivity is the parameter which largely affects the energy transfer rate but few studied are devoted towards the analysis of changes in density and specific heat. In the present study thermal conductivity model, specific heat model and density model are considered and analyzed for different nanocomposites. Different models are considered to validate the results and to reach a factual conclusion. Discussion is done in three different sections, in the first section, the effective thermal conductivity of the nanocomposites are analyzed, in second section effective density of the different nanocomposites are analyzed and in the third section, effective specific heat of the different nanocomposites are analyzed. In all the sections different mathematical models are utilized to analyse the effect of reinforcement of nanoparticle. Effect of increment in the nanoparticle concentration on the effective properties is also analyzed in all three sections.

RESULTS AND DISCUSSION

Thermal Conductivity Analysis

To analyze the application of nanocomposites in the field of energy, thermal conductivity, specific heat and density are studied. Four thermal conductivity models considered in the present study are based on, amount of concentration or volume fraction reinforcement of nanoparticle (Maxwell, 1873), thickness of the nanoparticle layers (Yu & Choi, 2003), nanoparticle shape factor (Hamilton & Crosser, 1962) and physical mixture rule (Rohatagi, 1993). Different values of thickness ratio (β) and shape factor/sphericity (n) is also varied to analyze their effects on the effective thermal conductivity. Six nanocomposites and three base materials as shown in table 1 are considered.

Table 1. Properties of nanoparticles and base materials

Material	Thermal Conductivity (W/m-K)	Specific Heat (J/kg-K)	Density (kg/m³)
B_4C	42	1288	2550
Al_2O_3	36	773	3880
SiC	100	1300	3200
TiO_2	8.9538	686.2	4250
ZnO	23.4	494	5675
SiO_2	1.5	730	2650
Al6061	167	896	2700
Al7075	130	960	2810
H_2O	0.6	4178	997.1

Mixture Rule

Physical mixture rule (Rohatagi, 1993) is also utilized to calculate these properties, model includes the concentration of nanocomposite and is expressed as,

$$k_{comb} = k_{comp}\phi + k_{bm}\left(1-\phi\right)$$

Maxwell model

$$k_{comb} = \left[\frac{\left(k_{comp} + 2k_{bm}\right) + 2\phi\left(k_{comp} - k_{bm}\right)}{\left(k_{comp} + 2k_{bm}\right) - \phi\left(k_{comp} - k_{bm}\right)}\right] k_{bf}$$

Yu & Choi model

$$k_{comb} = \left[\frac{\left(k_{comp} + 2k_{bm}\right) + 2\phi\left(k_{comp} - k_{bm}\right)\left(1+\beta\right)^{3}}{\left(k_{comp} + 2k_{bm}\right) - \phi\left(k_{comp} - k_{bm}\right)\left(1+\beta\right)^{3}}\right] k_{bf}$$

Hamilton & crosser model

$$k_{comb} = \left[\frac{k_{comp} + \left(n-1\right)k_{bm} + \left(n-1\right)\left(k_{comp} - k_{bm}\right)\phi}{k_{comp} + \left(n-1\right)k_{bm} - \left(k_{comp} - k_{bm}\right)\phi}\right] k_{bf}$$

In the above equations subscript comb, comp and bm represent the nanocomposites, nanoparticle and base material respectively. β is the nanolayer thickness to nanoparticle radius ratio and n is the shape factor of the nanoparticle (Singh, 2017).

Wide verities of nanoparticles considered in the present work are SiC, Al2O3, B4C, TiO_2, ZnO, and SiO_2, these nanoparticles are considered as they have good thermal conductivity. The concentration of these nanoparticles is varied from 0 to 30%. Three base materials on which these nanoparticles are reinforced are Al6061, Al7075 and H_2O. Tables 2, 3, 4, 5, 6 and 7 and figures 1, 2, 3, 4, 5 and 6 show the variation of thermal conductivity of Al6061-Al_2O_3, Al6061-B_4C, Al6061-SiC, Al6061-SiO_2, Al6061-TiO_2 and Al6061-ZnO with concentration. Tables 8-13 and figures 7-12 show the variation of thermal conductivity of Al7075-Al_2O_3, Al7075-B_4C, Al7075-SiC, Al7075-SiO_2, Al7075-TiO_2 and Al7075-ZnO with concentration.

In the below tables Y & C stands for Yu & Choi and H & C stands for Hamilton & Crosser. While ϕ represents the concentration percentage of nanocomposites reinforced into the base material.

Al6061

From figures 1, 2, 3, 4, 5 and 6 it can be seen that thermal conductivity for all the nanocomposites decreases with concentration. This decrement in the thermal conductivity of all the nanocomposites is can be due to the fact that thermal conductivity of all the six nanoparticles (SiC, Al2O3, B4C, TiO_2, ZnO and SiO_2) is less than the thermal conductivity of the base material Al6061. It can be seen from the figures 1, 2, 3, 4, 5 and 6 that Mixture rule shows a minimum decrement in the thermal

Table 2. Thermal conductivity vs concentration for Al6061-Al_2O_3

Al6061-Al_2O_3								
Φ	Mixture	Maxwell	Y & C β=0	Y & C β=0.2	Y & C β=0.3	H & C n=3	H & C n=4.9	H & C n=5.7
0	167	167	167	167	167	167	167	167
0.05	160.45	158.2852	158.2852	152.1292	143.7907	158.2852	159.2752	159.4649
0.1	153.9	149.8684	149.8684	138.1157	122.6366	149.8684	151.6948	152.0481
0.15	147.35	141.7346	141.7346	124.8875	103.2763	141.7346	144.255	144.7469
0.2	140.8	133.8698	133.8698	112.3804	85.49088	133.8698	136.9517	137.5586
0.25	134.25	126.2607	126.2607	100.537	69.09568	126.2607	129.7813	130.4807
0.3	127.7	118.8952	118.8952	89.30577	53.93385	118.8952	122.7402	123.5105

Table 3. Thermal conductivity vs concentration for Al6061- B_4C

Al6061-B_4C								
Φ	Mixture	Maxwell	Y & C β=0	Y & C β=0.2	Y & C β=0.3	H & C n=3	H & C n=4.9	H & C n=5.7
0	167	167	167	167	167	167	167	167
0.05	160.75	158.8084	158.8084	153.0114	145.1454	158.8084	159.6891	159.8592
0.1	154.5	150.8803	150.8803	139.7827	125.1177	150.8803	152.5076	152.8247
0.15	148.25	143.2033	143.2033	127.2537	106.6972	143.2033	145.4522	145.8942
0.2	142	135.7656	135.7656	115.3702	89.69771	135.7656	138.5196	139.0654
0.25	135.75	128.5562	128.5562	104.0837	73.96102	128.5562	131.7065	132.336
0.3	129.5	121.5647	121.5647	93.35027	59.35143	121.5647	125.0101	125.704

Table 4. Thermal conductivity vs concentration for Al6061- SiC

Al6061-SiC								
Φ	Mixture	Maxwell	Y & C β=0	Y & C β=0.2	Y & C β=0.3	H & C n=3	H & C n=4.9	H & C n=5.7
0	167	167	167	167	167	167	167	167
0.05	163.65	163.1625	163.1625	160.4055	156.6086	163.1625	163.3674	163.4099
0.1	160.3	159.3833	159.3833	153.9823	146.6395	159.3833	159.767	159.8469
0.15	156.95	155.6611	155.6611	147.7239	137.0675	155.6611	156.1982	156.3105
0.2	153.6	151.9946	151.9946	141.624	127.8692	151.9946	152.6608	152.8004
0.25	150.25	148.3827	148.3827	135.6767	119.0234	148.3827	149.1541	149.3165
0.3	146.9	144.824	144.824	129.8762	110.51	144.824	145.6779	145.8584

Table 5. Thermal conductivity vs concentration for Al6061- SiO_2

Al6061-SiO_2								
Φ	Mixture	Maxwell	Y & C β=0	Y & C β=0.2	Y & C β=0.3	H & C n=3	H & C n=4.9	H & C n=5.7
0	167	167	167	167	167	167	167	167
0.05	158.725	154.9404	154.9404	146.52	135.2418	154.9404	156.7569	157.0878
0.1	150.45	143.4478	143.4478	127.6486	107.2698	143.4478	146.7671	147.3799
0.15	142.175	132.483	132.483	110.2034	82.44513	132.483	137.0213	137.8701
0.2	133.9	122.0106	122.0106	94.0287	60.26482	122.0106	127.5107	128.5523
0.25	125.625	111.998	111.998	78.99055	40.32772	111.998	118.2268	119.4208
0.3	117.35	102.4157	102.4157	64.97321	22.30992	102.4157	109.1616	110.4701

Table 6. Thermal conductivity vs concentration for Al6061- TiO_2

Al6061-TiO_2								
Φ	Mixture	Maxwell	Y & C β=0	Y & C β=0.2	Y & C β=0.3	H & C n=3	H & C n=4.9	H & C n=5.7
0	167	167	167	167	167	167	167	167
0.05	159.0977	155.716	155.716	147.8158	137.207	155.716	157.3219	157.6178
0.1	151.1954	144.9291	144.9291	130.0466	110.7585	144.9291	147.8701	148.4188
0.15	143.2931	134.6072	134.6072	113.5415	87.1213	134.6072	138.6367	139.3976
0.2	135.3908	124.7208	124.7208	98.17013	65.86975	124.7208	129.6142	130.5491
0.25	127.4885	115.2429	115.2429	83.8196	46.66004	115.2429	120.7955	121.8685
0.3	119.5861	106.1488	106.1488	70.39148	29.2114	106.1488	112.1738	113.3509

Table 7. Thermal conductivity vs concentration for Al6061- ZnO

Al6061-ZnO								
Φ	**Mixture**	**Maxwell**	**Y & C β=0**	**Y & C β=0.2**	**Y & C β=0.3**	**H & C n=3**	**H & C n=4.9**	**H & C n=5.7**
0	167	167	167	167	167	167	167	167
0.05	159.82	157.1334	157.1334	150.1914	140.8249	157.1334	158.3835	158.6189
0.1	152.64	147.6478	147.6478	134.4741	117.2492	147.6478	149.9466	150.384
0.15	145.46	138.5218	138.5218	119.7451	95.90378	138.5218	141.6838	142.2917
0.2	138.28	129.7351	129.7351	105.914	76.48658	129.7351	133.5896	134.3382
0.25	131.1	121.2693	121.2693	92.90122	58.7476	121.2693	125.6589	126.52
0.3	123.92	113.107	113.107	80.63614	42.47827	113.107	117.887	118.8337

conductivity for all the nanocomposites considered. It can be seen that Yu & Choi model for β=0.3 shows the highest decrement in the thermal conductivity for all the nanocomposites considered. As β represents the ratio of nanolayer thickness to nanoparticle radius, a higher value of β means high nanolayer thickness that means lower the thermal conductivity. It can also be seen that for β=0 and n=3 Maxwell mode, Yu & Choi model and Hamilton & Crosser models overlap each other. From this analyses it can be concluded that reinforcing a nanoparticle with lesser thermal conductivity into a base material with higher thermal conductivity will results in the lesser effective thermal conductivity nanocomposites, while reinforcing a nanoparticle with higher thermal conductivity value into a base material with lower thermal conductivity will result in a nanocomposite having higher effective thermal conductivity (Sarvia & Fuskele, 2017). So from the above discussion, it can be said that to enhance the energy transfer it is better to reinforce a nanoparticle with higher thermal conductivity. These higher thermal conductivity nanocomposites are suitable for high heat releasing energy devices.

Figure 1. Thermal conductivity vs concentration variation Al6061-Al_2O_3

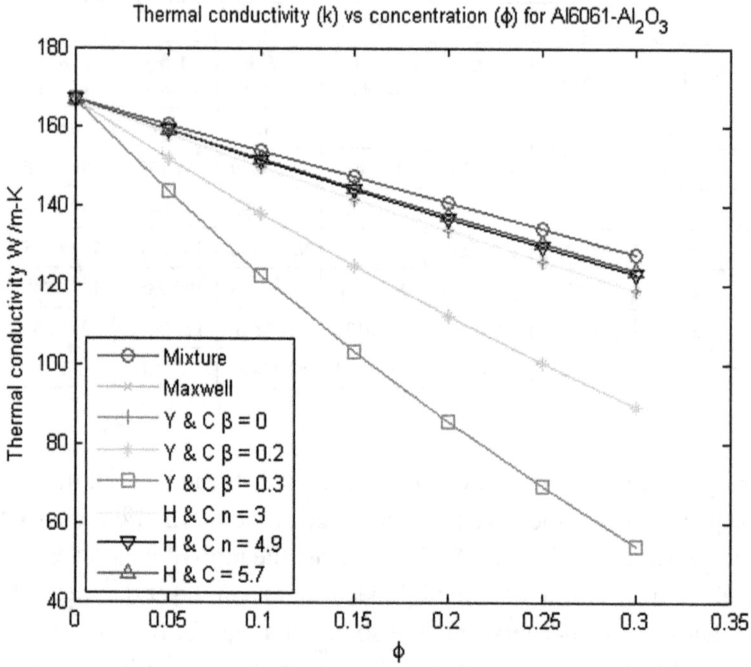

Figure 2. Thermal conductivity vs concentration variation Al6061-B_4C

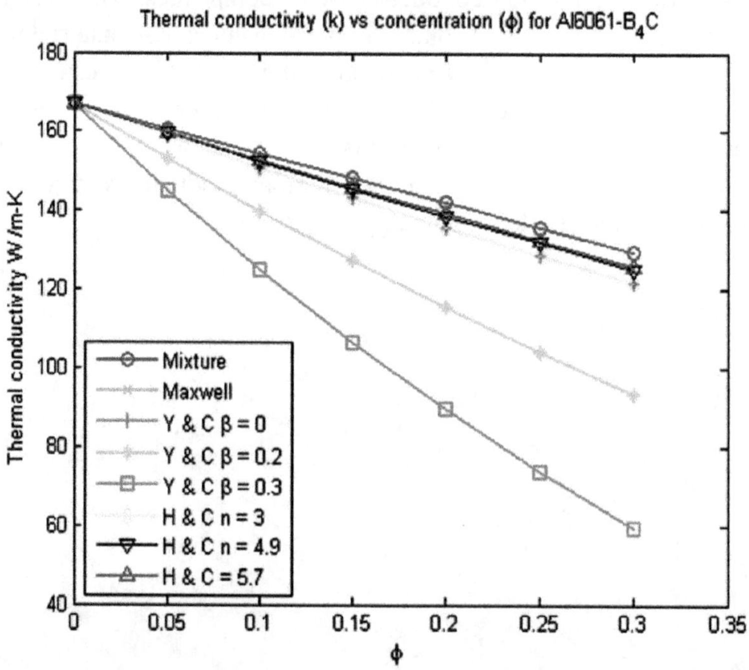

Figure 3. Thermal conductivity vs concentration variation Al6061-SiC

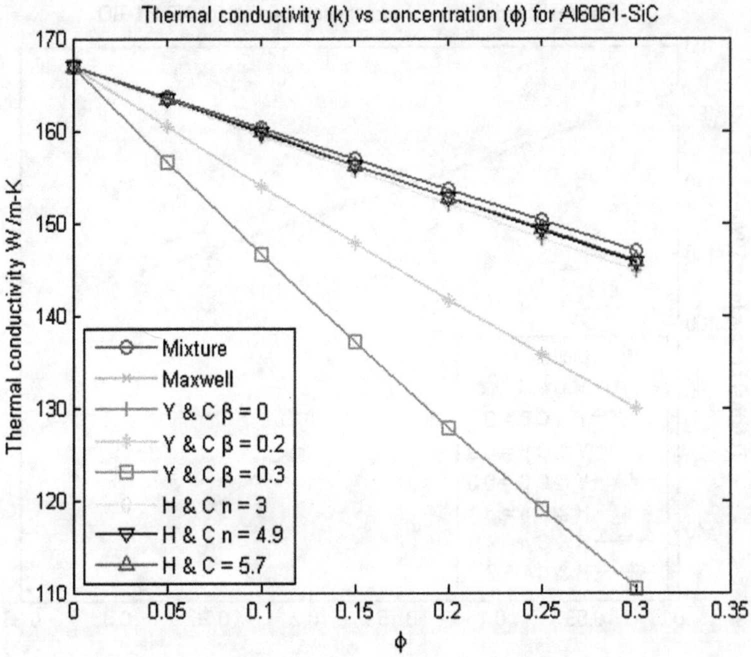

Figure 4. Thermal conductivity vs concentration variation Al6061-SiO_2

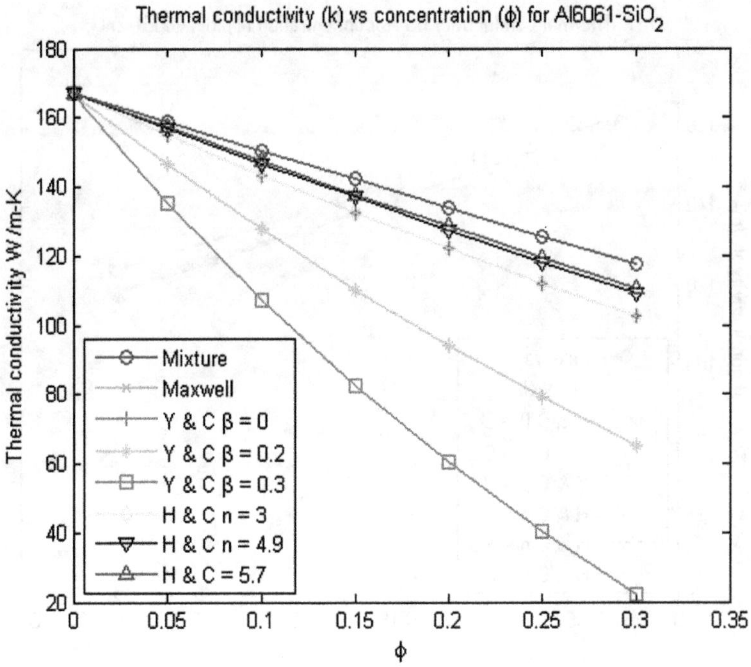

Figure 5. Thermal conductivity vs concentration variation Al6061-TiO_2

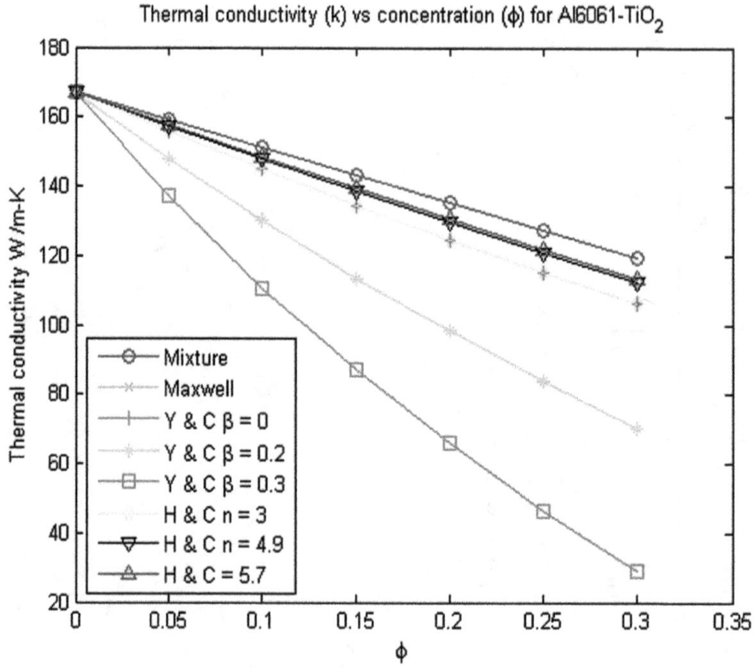

Figure 6. Thermal conductivity vs concentration variation Al6061-ZnO

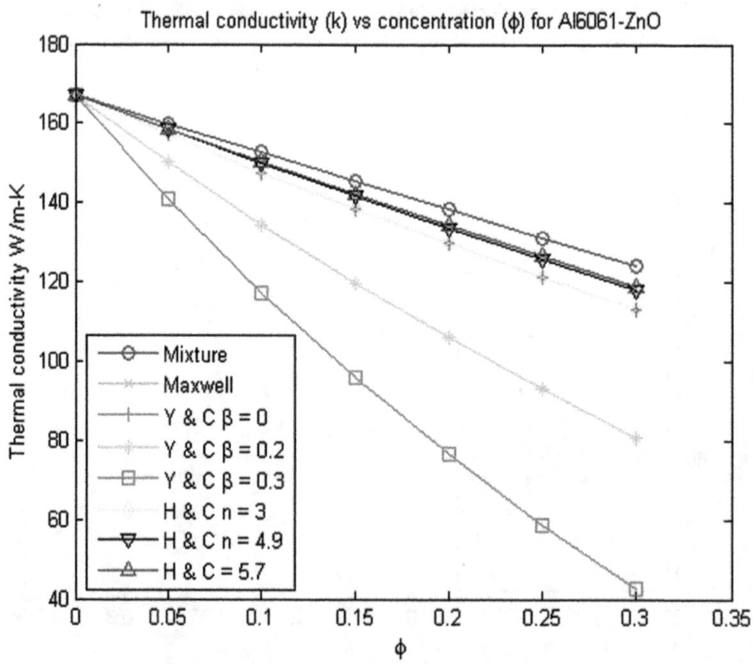

Table 8. Thermal conductivity vs concentration for Al7075- Al_2O_3

Al7075-Al_2O_3								
Φ	Mixture	Maxwell	Y & C β=0	Y & C β=0.2	Y & C β=0.3	H & C n=3	H & C n=4.9	H & C n=5.7
0	130	130	130	130	130	130	130	130
0.05	125.3	123.9042	123.9042	119.585	113.717	123.9042	124.5337	124.656
0.1	120.6	117.9961	117.9961	109.7118	98.73927	117.9961	119.1604	119.3885
0.15	115.9	112.267	112.267	100.3392	84.91578	112.267	113.8778	114.1959
0.2	111.2	106.709	106.709	91.43007	72.1181	106.709	108.6835	109.0766
0.25	106.5	101.3146	101.3146	82.95085	60.23613	101.3146	103.5755	104.0291
0.3	101.8	96.0765	96.0765	74.87116	49.17498	96.0765	98.55147	99.05184

Table 9. Thermal conductivity vs concentration for Al7075- B_4C

Al7075-B_4C								
Φ	Mixture	Maxwell	Y & C β=0	Y & C β=0.2	Y & C β=0.3	H & C n=3	H & C n=4.9	H & C n=5.7
0	130	130	130	130	130	130	130	130
0.05	125.6	124.3995	124.3995	120.4224	115.0076	124.3995	124.9353	125.0405
0.1	121.2	118.9575	118.9575	111.304	101.1253	118.9575	119.9505	120.1469
0.15	116.8	113.6675	113.6675	102.6124	88.23405	113.6675	115.0438	115.3179
0.2	112.4	108.5232	108.5232	94.31849	76.23147	108.5232	110.2132	110.5523
0.25	108	103.5185	103.5185	86.39546	65.0287	103.5185	105.4571	105.8489
0.3	103.6	98.64799	98.64799	78.81903	54.54838	98.64799	100.7737	101.2064

Table 10. Thermal conductivity vs concentration for Al7075- SiC

Al7075-SiC								
Φ	Mixture	Maxwell	Y & C β=0	Y & C β=0.2	Y & C β=0.3	H & C n=3	H & C n=4.9	H & C n=5.7
0	130	130	130	130	130	130	130	130
0.05	128.5	128.3817	128.3817	127.2121	125.5914	128.3817	128.4297	128.44
0.1	127	126.7769	126.7769	124.4637	121.2814	126.7769	126.8672	126.8866
0.15	125.5	125.1852	125.1852	121.7541	117.0666	125.1852	125.3123	125.3396
0.2	124	123.6066	123.6066	119.0824	112.944	123.6066	123.7651	123.7992
0.25	122.5	122.0408	122.0408	116.4479	108.9106	122.0408	122.2254	122.2651
0.3	121	120.4878	120.4878	113.8497	104.9635	120.4878	120.6932	120.7375

Table 11. Thermal conductivity vs concentration for Al7075- SiO_2

Al7075-SiO_2								
Φ	**Mixture**	**Maxwell**	**Y & C β=0**	**Y & C β=0.2**	**Y & C β=0.3**	**H & C n=3**	**H & C n=4.9**	**H & C n=5.7**
0	130	130	130	130	130	130	130	130
0.05	123.575	120.6476	120.6476	114.1163	105.3671	120.6476	122.0518	122.3078
0.1	117.15	111.7332	111.7332	99.47576	83.66108	111.7332	114.2995	114.7736
0.15	110.725	103.2268	103.2268	85.93796	64.38947	103.2268	106.736	107.3926
0.2	104.3	95.10097	95.10097	73.3828	47.16449	95.10097	99.35436	100.1603
0.25	97.875	87.33078	87.33078	61.70706	31.67656	87.33078	92.14821	93.07213
0.3	91.45	79.89335	79.89335	50.82147	17.67552	79.89335	85.11132	86.12388

Table 12. Thermal conductivity vs concentration for Al7075- TiO_2

Al7075-TiO_2								
Φ	**Mixture**	**Maxwell**	**Y & C β=0**	**Y & C β=0.2**	**Y & C β=0.3**	**H & C n=3**	**H & C n=4.9**	**H & C n=5.7**
0	130	130	130	130	130	130	130	130
0.05	123.9477	121.4169	121.4169	115.4023	107.3186	121.4169	122.6144	122.8359
0.1	117.8954	113.2035	113.2035	101.858	87.13033	113.2035	115.3981	115.809
0.15	111.8431	105.3363	105.3363	89.25695	69.04562	105.3363	108.3454	108.9154
0.2	105.7908	97.79402	97.79402	77.50401	52.75194	97.79402	101.4507	102.1514
0.25	99.73845	90.55682	90.55682	66.51632	37.99579	90.55682	94.70878	95.51326
0.3	93.68614	83.6066	83.6066	56.22149	24.56934	83.6066	88.11464	88.99757

Table 13. Thermal conductivity vs concentration for Al7075- ZnO

Al7075-ZnO								
Φ	**Mixture**	**Maxwell**	**Y & C β=0**	**Y & C β=0.2**	**Y & C β=0.3**	**H & C n=3**	**H & C n=4.9**	**H & C n=5.7**
0	130	130	130	130	130	130	130	130
0.05	124.67	122.8005	122.8005	117.7243	110.8609	122.8005	123.6625	123.8263
0.1	119.34	115.8621	115.8621	106.1978	93.51234	115.8621	117.4498	117.7545
0.15	114.01	109.1706	109.1706	95.35397	77.71438	109.1706	111.3583	111.7823
0.2	108.68	102.7133	102.7133	85.13386	63.268	102.7133	105.3845	105.9072
0.25	103.35	96.47799	96.47799	75.48519	50.00686	96.47799	99.52509	100.1268
0.3	98.02	90.45342	90.45342	66.36134	37.79084	90.45342	93.7767	94.43894

Al7075

From the tables 8, 9, 10, 11, 12 and 13 and figures 7, 8, 9, 10, 11 and 12 it can be seen that thermal conductivity for all the nanocomposites decreases with concentration. This decrement in the thermal conductivity of all the nanocomposites is can be due to the fact that thermal conductivity of all the six nanoparticles (SiC, Al2O3, B4C, TiO_2, ZnO and SiO_2) is less than the thermal conductivity of the base material Al7075. It can be seen from the figures 7, 8, 9, 10, 11 and 12 that Mixture rule shows a minimum decrement in the thermal conductivity for all the nanocomposites considered. It can be seen that Yu & Choi model for β=0.3 shows the highest decrement in the thermal conductivity for all the nanocomposites considered. As β represents the ratio of nanolayer thickness to nanoparticle radius, a higher value of β means high nanolayer thickness at the top surface that means lower the thermal conductivity. It can also be seen that for β=0 and n=3 Maxwell mode, Yu & Choi model and Hamilton & Crosser models overlap each other which is due to the fact that for β=0 and n=3 Yu & Choi model and Hamilton & Crosser model coverts to Maxwell model. From this analyses it can be concluded that reinforcing a nanoparticle with lesser thermal conductivity into a base material with higher thermal conductivity will results in the lesser effective thermal conductivity nanocomposites, while reinforcing a nanoparticle with higher thermal conductivity value into a base material with lower thermal conductivity compared to the reinforced nanoparticle, will result in a nanocomposite having higher effective thermal conductivity (Sarvia & Fuskele, 2017). So from the above discussion, it can be said that to enhance the energy transfer it is better to reinforce a nanoparticle with higher thermal conductivity.

H_2O

From figures 13, 14, 15, 16, 17 and 18 it can be seen that thermal conductivity for all the nanocomposites increases. It can be seen from the figure 13, 14, 15, 16, 17 and 18 that Mixture rule shows maximum increment in the thermal conductivity for all the nanocomposites considered, except the H_2O-TiO_2 and H_2O-SiO_2 nanocomposites for highest values of concentration. This increment in the thermal conductivity of all the nanocomposites can be due to the fact that thermal conductivity of all the six nanoparticles (SiC, Al2O3, B4C, TiO_2, ZnO and SiO_2) is higher than the thermal conductivity of the base material (H_2O). So reinforcing a nanoparticle with higher thermal conductivity into a base material with lesser thermal conductivity will result in the higher effective thermal conductivity (Sarvia & Fuskele, 2017).

From the above discussion, it can be concluded that mixing a low thermal conductivity value nanoparticle into a higher thermal conductivity base material will result in lower effective thermal conductivity nanocomposites. These types

Figure 7. Thermal conductivity vs concentration variation Al7075-Al_2O_3

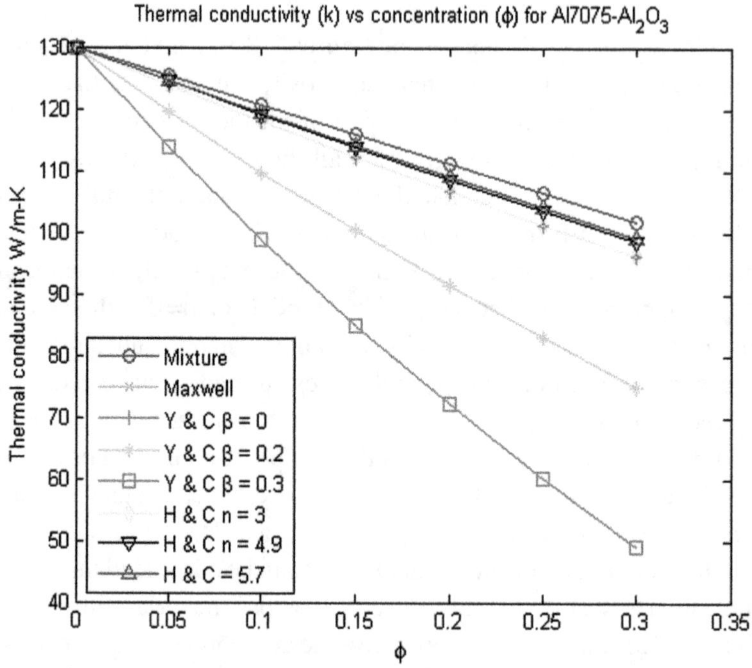

Figure 8. Thermal conductivity vs concentration variation Al7075-B_4C

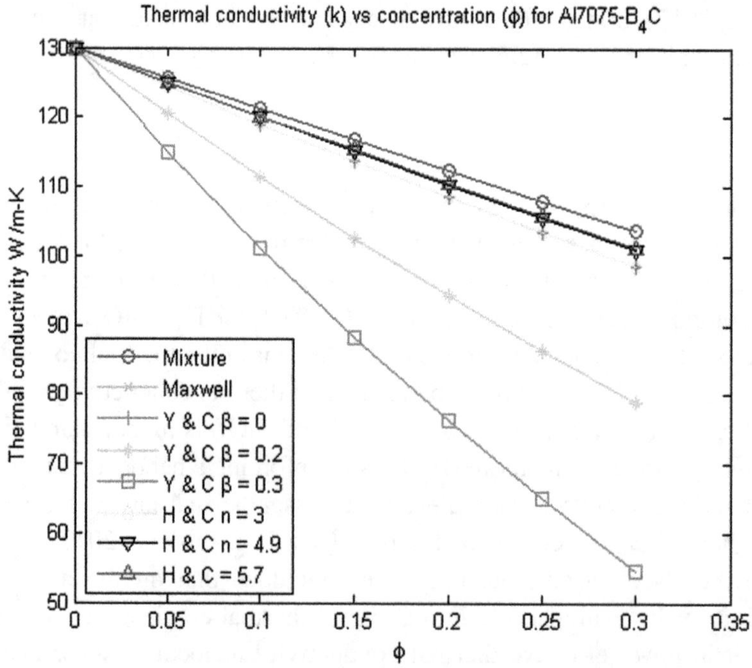

Figure 9. Thermal conductivity vs concentration variation Al7075-SiC

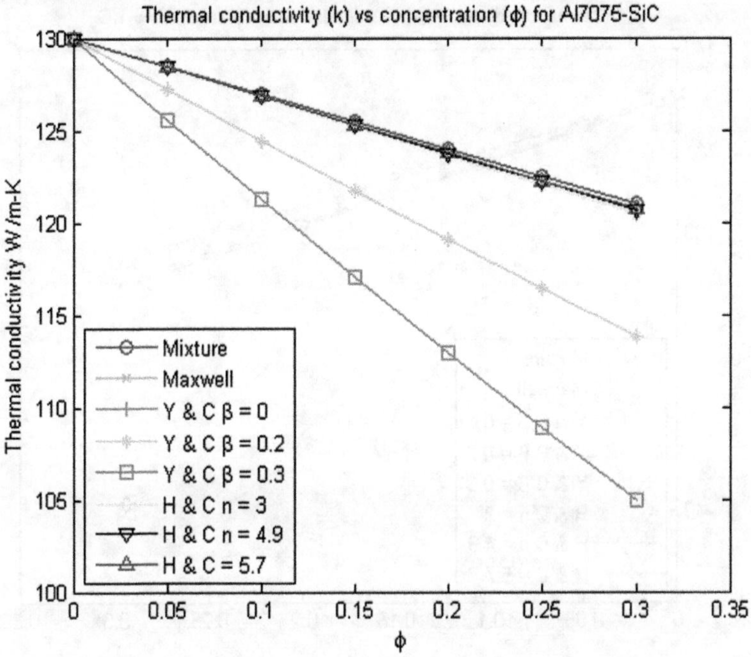

Figure 10. Thermal conductivity vs concentration variation Al7075-SiO_2

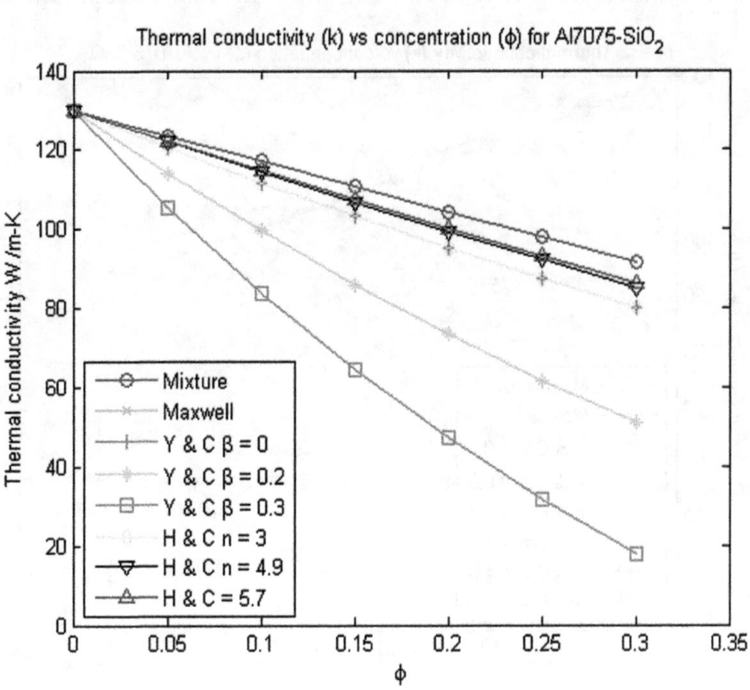

Figure 11. Thermal conductivity vs concentration variation Al7075-TiO_2

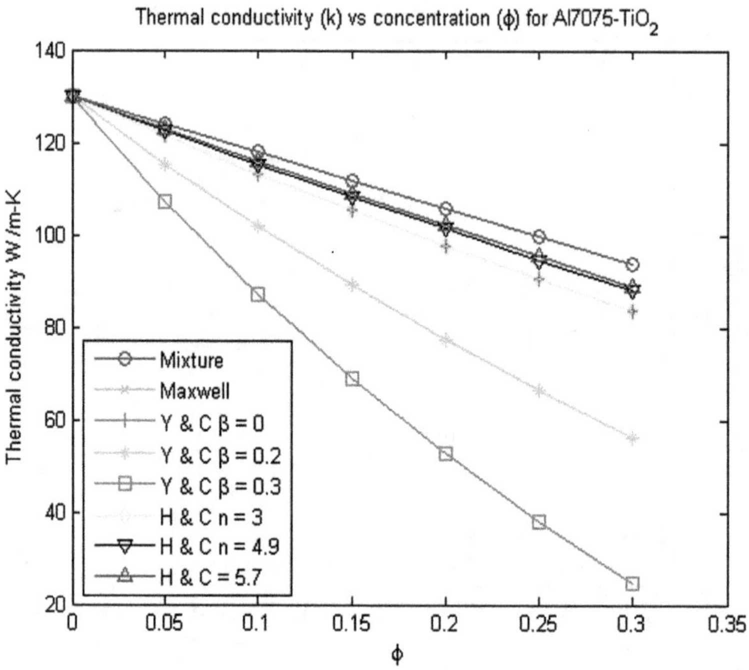

Figure 12. Thermal conductivity vs concentration variation Al7075-ZnO

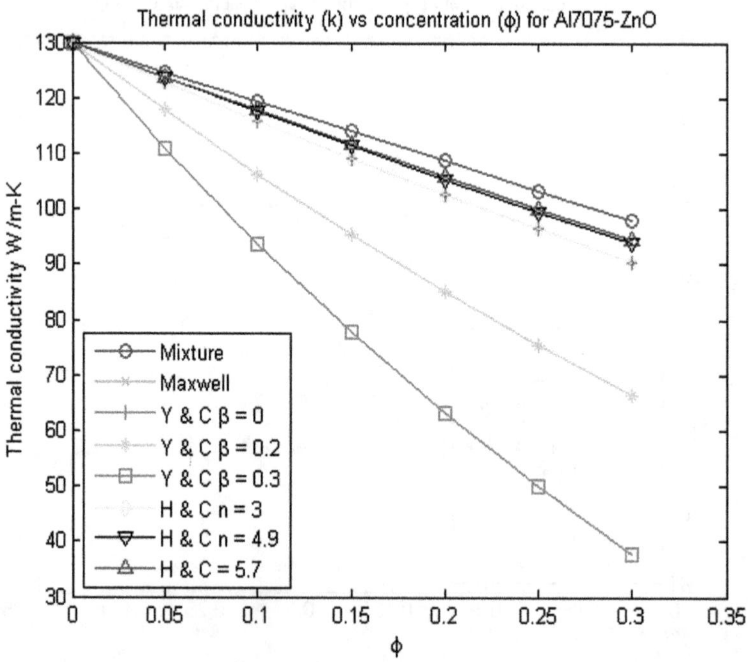

Table 14. Thermal conductivity vs concentration for H_2O- Al_2O_3

H_2O-Al_2O_3								
Φ	**Mixture**	**Maxwell**	**Y & C β=0**	**Y & C β=0.2**	**Y & C β=0.3**	**H & C n=3**	**H & C n=4.9**	**H & C n=5.7**
0	0.6000	0.6000	0.6000	0.6000	0.6000	0.6000	0.6000	0.6000
0.05	2.3700	0.6899	0.6899	0.7613	0.8703	0.6899	0.7423	0.7634
0.1	4.1400	0.7893	0.7893	0.9542	1.2361	0.7893	0.8991	0.9432
0.15	5.9100	0.8997	0.8997	1.1894	1.7590	0.8997	1.0726	1.1419
0.2	7.6800	1.0231	1.0231	1.4821	2.5676	1.0231	1.2659	1.3629
0.25	9.4500	1.1619	1.1619	1.8565	3.9844	1.1619	1.4823	1.6099
0.3	11.2200	1.3192	1.3192	2.3525	7.1090	1.3192	1.7264	1.8880

Table 15. Thermal conductivity vs concentration for H_2O- B_4C

H_2O-B_4C								
Φ	**Mixture**	**Maxwell**	**Y & C β=0**	**Y & C β=0.2**	**Y & C β=0.3**	**H & C n=3**	**H & C n=4.9**	**H & C n=5.7**
0	0.6000	0.6000	0.6000	0.6000	0.6000	0.6000	0.6000	0.6000
0.05	2.6700	0.6906	0.6906	0.7625	0.8725	0.6906	0.7440	0.7656
0.1	4.7400	0.7908	0.7908	0.9572	1.2422	0.7908	0.9028	0.9481
0.15	6.8100	0.9022	0.9022	1.1949	1.7725	0.9022	1.0788	1.1501
0.2	8.8800	1.0268	1.0268	1.4914	2.5969	1.0268	1.2751	1.3750
0.25	10.9500	1.1671	1.1671	1.8717	4.0542	1.1671	1.4952	1.6269
0.3	13.0200	1.3263	1.3263	2.3771	7.3268	1.3263	1.7439	1.9110

Table 16. Thermal conductivity vs concentration for H_2O- SiC

H_2O-SiC								
Φ	**Mixture**	**Maxwell**	**Y & C β=0**	**Y & C β=0.2**	**Y & C β=0.3**	**H & C n=3**	**H & C n=4.9**	**H & C n=5.7**
0	0.6000	0.6000	0.6000	0.6000	0.6000	0.6000	0.6000	0.6000
0.05	5.5700	0.6930	0.6930	0.7669	0.8803	0.6930	0.7501	0.7737
0.1	10.5400	0.7961	0.7961	0.9680	1.2641	0.7961	0.9163	0.9660
0.15	15.5100	0.9110	0.9110	1.2148	1.8215	0.9110	1.1014	1.1801
0.2	20.4800	1.0400	1.0400	1.5250	2.7049	1.0400	1.3088	1.4197
0.25	25.4500	1.1859	1.1859	1.9267	4.3181	1.1859	1.5428	1.6900
0.3	30.4200	1.3520	1.3520	2.4673	8.2023	1.3520	1.8089	1.9970

Table 17. Thermal conductivity vs concentration for H_2O- SiO_2

H_2O-SiO_2								
Φ	Mixture	Maxwell	Y & C β=0	Y & C β=0.2	Y & C β=0.3	H & C n=3	H & C n=4.9	H & C n=5.7
0	0.6000	0.6000	0.6000	0.6000	0.6000	0.6000	0.6000	0.6000
0.05	0.6450	0.6305	0.6305	0.6534	0.6863	0.6305	0.6349	0.6360
0.1	0.6900	0.6621	0.6621	0.7100	0.7812	0.6621	0.6706	0.6728
0.15	0.7350	0.6947	0.6947	0.7702	0.8862	0.6947	0.7071	0.7103
0.2	0.7800	0.7286	0.7286	0.8344	1.0030	0.7286	0.7446	0.7487
0.25	0.8250	0.7636	0.7636	0.9028	1.1336	0.7636	0.7830	0.7879
0.3	0.8700	0.8000	0.8000	0.9760	1.2807	0.8000	0.8224	0.8280

Table 18. Thermal conductivity vs concentration for H_2O- TiO_2

H_2O-TiO_2								
Φ	Mixture	Maxwell	Y & C β=0	Y & C β=0.2	Y & C β=0.3	H & C n=3	H & C n=4.9	H & C n=5.7
0	0.6000	0.6000	0.6000	0.6000	0.6000	0.6000	0.6000	0.6000
0.05	1.0177	0.6772	0.6772	0.7377	0.8290	0.6772	0.7129	0.7258
0.1	1.4354	0.7614	0.7614	0.8983	1.1248	0.7614	0.8348	0.8612
0.15	1.8531	0.8534	0.8534	1.0879	1.5216	0.8534	0.9669	1.0073
0.2	2.2708	0.9545	0.9545	1.3151	2.0818	0.9545	1.1104	1.1656
0.25	2.6885	1.0661	1.0661	1.5925	2.9321	1.0661	1.2670	1.3375
0.3	3.1061	1.1899	1.1899	1.9386	4.3774	1.1899	1.4385	1.5248

Table 19. Thermal conductivity vs concentration for H_2O- ZnO

H_2O-ZnO								
Φ	Mixture	Maxwell	Y & C β=0	Y & C β=0.2	Y & C β=0.3	H & C n=3	H & C n=4.9	H & C n=5.7
0	0.6000	0.6000	0.6000	0.6000	0.6000	0.6000	0.6000	0.6000
0.05	1.7400	0.6875	0.6875	0.7567	0.8622	0.6875	0.7362	0.7555
0.1	2.8800	0.7839	0.7839	0.9433	1.2139	0.7839	0.8857	0.9257
0.15	4.0200	0.8907	0.8907	1.1692	1.7102	0.8907	1.0505	1.1130
0.2	5.1600	1.0096	1.0096	1.4483	2.4633	1.0096	1.2330	1.3200
0.25	6.3000	1.1429	1.1429	1.8020	3.7424	1.1429	1.4362	1.5500
0.3	7.4400	1.2932	1.2932	2.2647	6.3938	1.2932	1.6640	1.8071

Figure 13. Thermal conductivity vs concentration variation H_2O-Al_2O_3

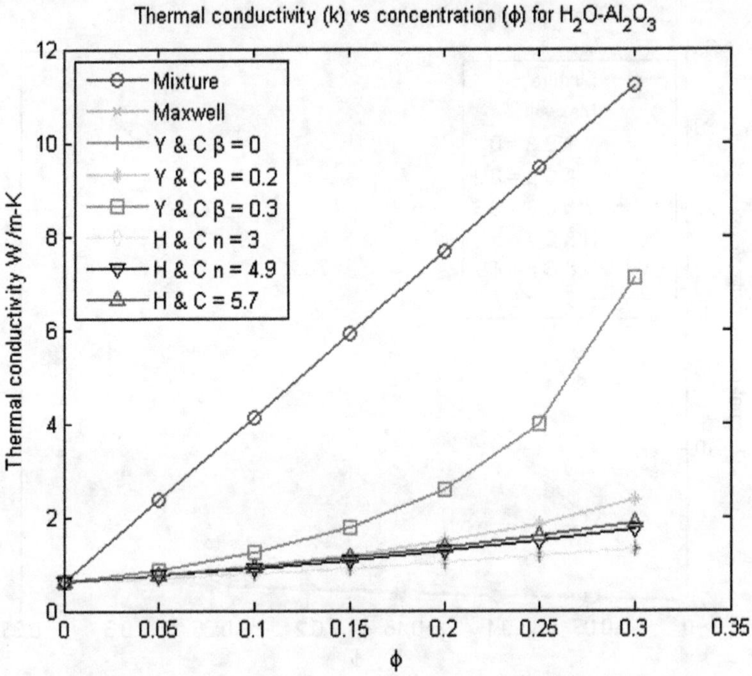

Figure 14. Thermal conductivity vs concentration variation H_2O-B_4C

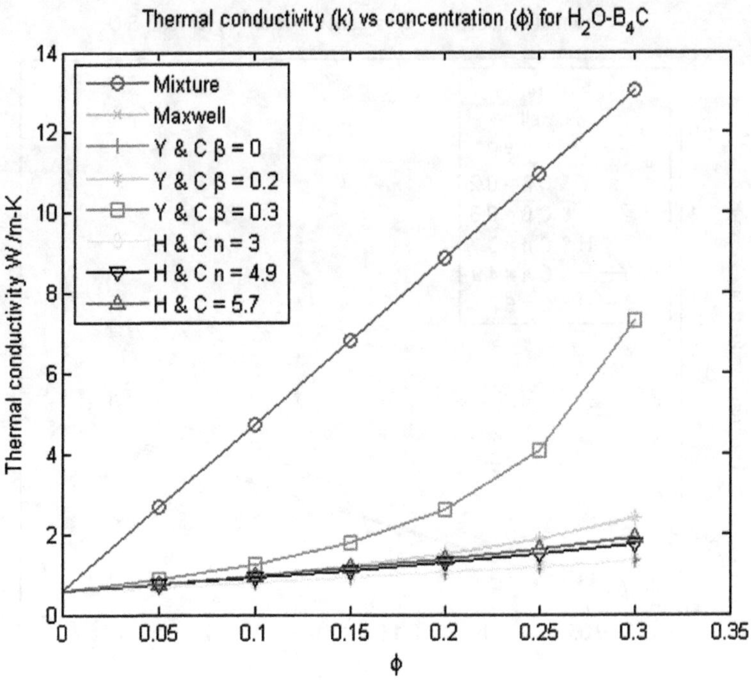

Figure 15. Thermal conductivity vs concentration variation H_2O-SiC

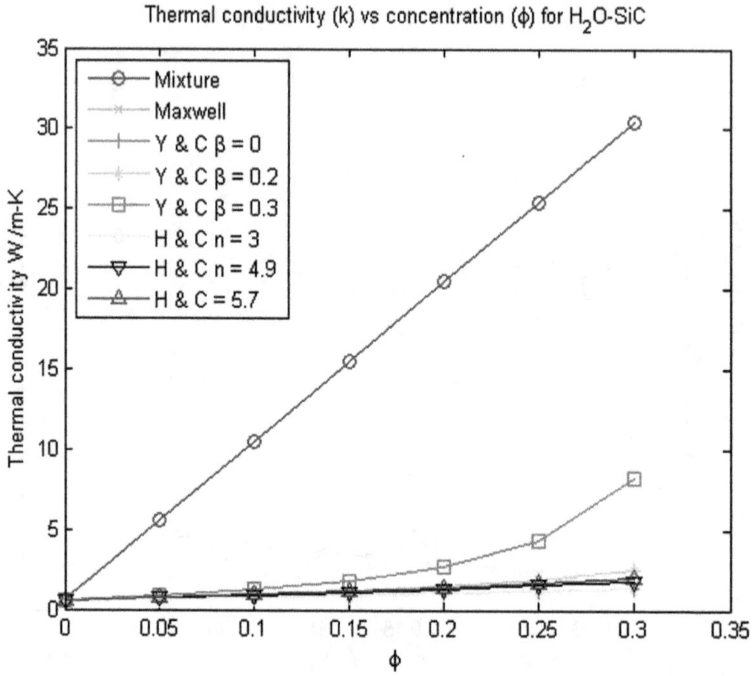

Figure 16. Thermal conductivity vs concentration variation H_2O-SiO_2

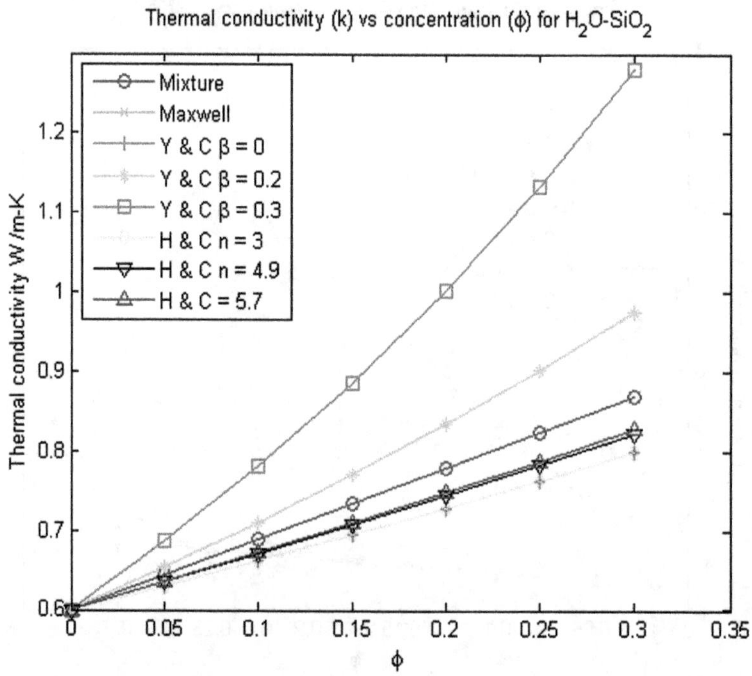

Figure 17. Thermal conductivity vs concentration variation H_2O-TiO_2

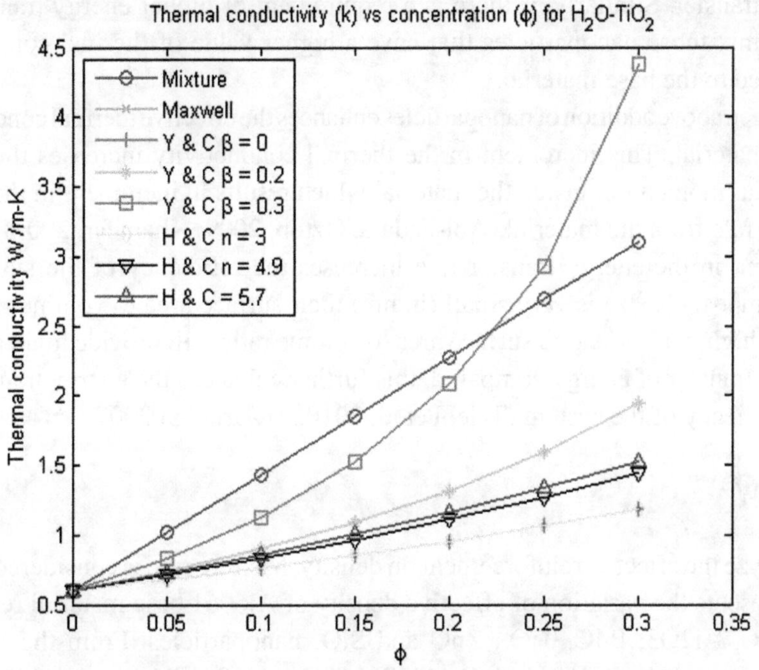

Figure 18. Thermal conductivity vs concentration variation H_2O-ZnO

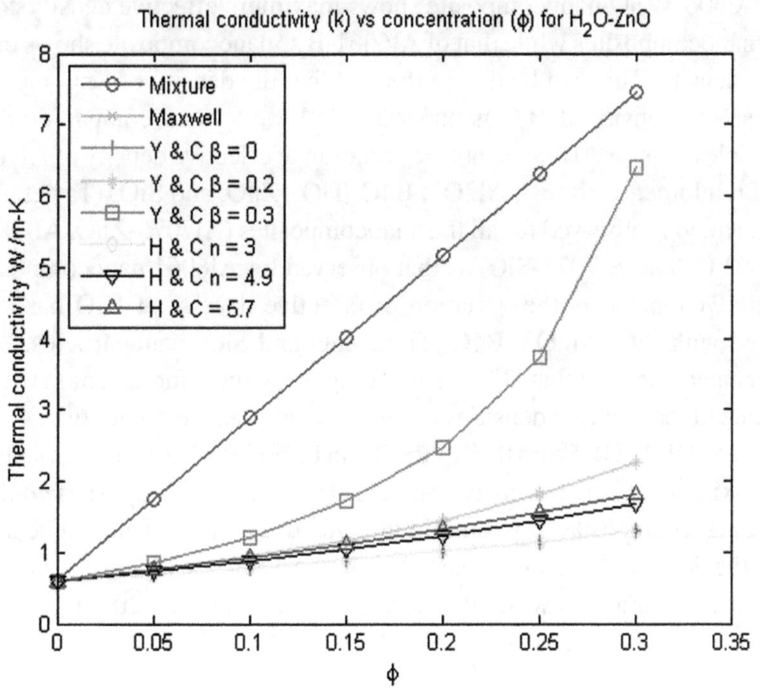

of nanocomposites are suitable where there is a requirement of insulation or less energy transfer. Similarly, if there is a requirement of higher energy transfer one should mix those nanoparticles that have a higher value of thermal conductivity compared to the base material.

As seen above addition of nanoparticles enhances the effective thermal conductivity of the material. This increment in the thermal conductivity increases the rate of molecular momentum inside the material which results in increment in the energy transfer rate from the material (Abi-nada & Oztop, 2009; Khanafer, 2007) and this increment in the energy transfer rate increases the efficiency of the system. As nanocomposite's size is very small (in nm) their surface area to volume ratio will be very high, this increased surface area to volume ratio will provide higher surface area for transfer of energy compared, this further enhances the energy transfer rate and efficiency of the system (Talebi et al, 2010; Tiwari et al 2007).

Density

To analyze the effect of reinforcement on density, Mixture rule is considered. Figure 19 represents the variation of effective density of Al6061 base material reinforced with SiC, Al2O3, B4C, TiO_2, ZnO and SiO_2 nanoparticles. From the figure, it can be noticed that Al6061-ZnO, Al6061-TiO_2, Al6061-Al_2O_3, and Al6061-SiC nanocomposites show increment in the effective density while Al6061-SiO_2 and Al6061-B_4C show a decrement in the effective density. From the figure, it can be noticed Al6061-ZnO nanocomposite shows maximum effective density compared to other nanocomposites while that of Al6061-B_4C nanocomposite shows minimum effective density. This can be due to the fact that the density of ZnO nanoparticle is highest and density of B_4C nanoparticle is lowest when compared with other nanoparticles. Figure 20 represents the variation of effective density of Al7075 base material reinforced with SiC, Al2O3, B4C, TiO_2, ZnO, and SiO_2 (Tay et al, 2017). The same trend is observed for all the nanocomposites (Al7075-ZnO, Al7075-TiO_2, Al7075-Al_2O_3, and Al7075-SiC) as that observed for Al6061 nanocomposites.

Figure 21 represents the variation of effective density of H_2O base material reinforced with SiC, Al2O3, B4C, TiO_2, ZnO and SiO_2 nanoparticles. From the figure, it can be noticed that all the nanocomposites show increment in the density. This is due to the fact that densities of all nanoparticles are higher than the density of base material (H_2O). From the figure, it can be noticed H_2O-ZnO nanocomposite shows maximum effective density compared to other nanocomposites while that of H_2O-B_4C nanocomposite shows minimum effective density. This can be due to the fact that the density of ZnO nanoparticle is highest and density of B_4C nanoparticle is lowest when compared with other nanoparticles (Tay et al, 2017).

Figure 19. Density variations of Al6061-Al_2O_3, Al6061-B_4C, Al6061-SiC, Al6061-SiO_2, Al6061-TiO_2, and Al6061-ZnO

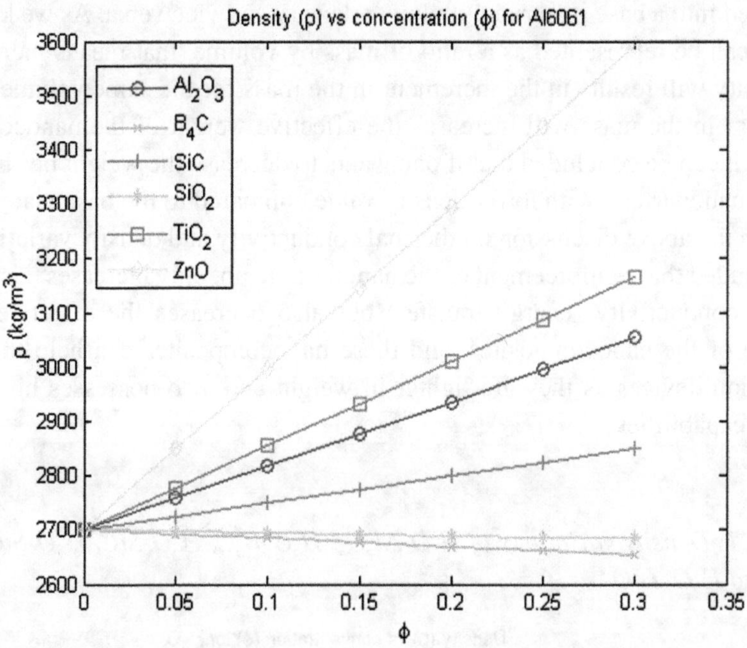

Figure 20. Density variations of Al7075-Al_2O_3, Al7075-B_4C, Al7075-SiC, Al7075-SiO_2, Al7075-TiO_2, and Al7075-ZnO

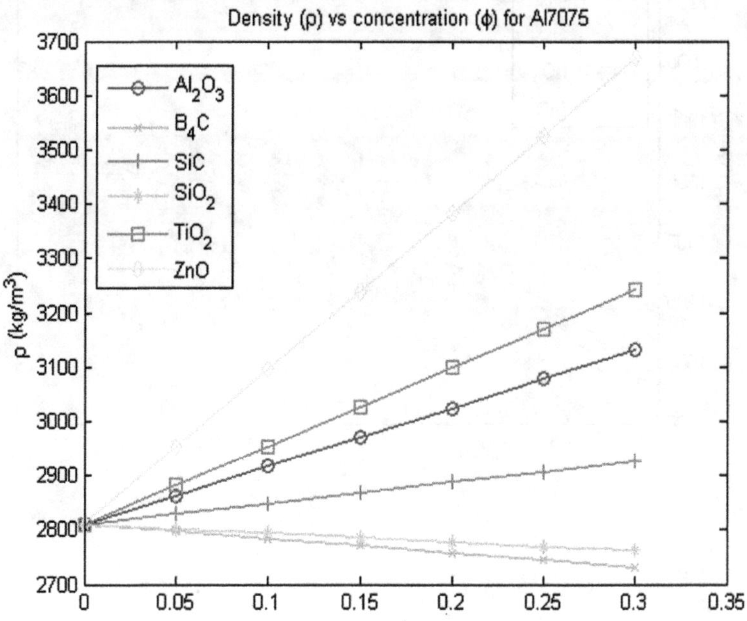

From the above discussion, it can be concluded that addition of nanoparticle having higher density increases the effective density of the nanocomposite when reinforced into a base material with lower density and vice versa. As we know that density can be represented as a ratio of mass by volume, that means increment in the density will results in the increment in the mass for the same volume and this increment in the mass will increases the effective weight of the nanocomposite. Further, it can be concluded that if one wants to decrease the weight he/she should mix the nanoparticle with lower density value compared to the base material.

From the above discussion on thermal conductivity and density variation it can be concluded that reinforcement of the nanoparticle not only increases the effective thermal conductivity (energy transfer) but also decreases the effective density (weight) of the nanocomposites, and these nanocomposites can help the energy application devices as they are lighter in weight and also possesses high energy transfer capabilities.

Figure 21. Density variations of H_2O-Al_2O_3, H_2O-B_4C, H_2O-SiC, H_2O-SiO_2, H_2O-TiO_2, and H_2O-ZnO

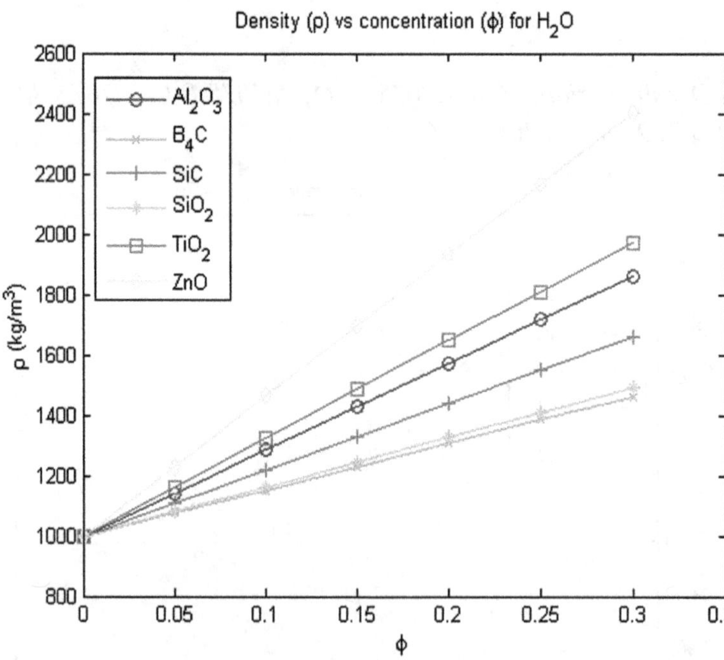

Specific Heat

To analyze the effect of reinforcement of nanoparticles on the base material specific heat, Mixture rule is considered. Figure 22 represents the variation of effective specific heat of Al6061 base material reinforced with SiC, Al2O3, B4C, TiO_2, ZnO and SiO_2 nanoparticles. From the figure, it can be noticed that Al6061-ZnO, Al6061-TiO_2, Al6061-Al_2O_3 and Al6061-SiO_2 nanocomposites show increment in the specific heat while that of Al6061-SiC and Al6061-B_4C shows a decrement in the specific heat. From the figure, it can be noticed Al6061-ZnO nanocomposite shows maximum effective specific heat compared to other nanocomposites while that of Al6061-SiC nanocomposite shows minimum specific heat. This can be due to the fact that specific heat of ZnO nanocomposite is highest and specific heat of SiC is lowest when compared with other nanocomposites. Figure 23 represents the variation of effective specific heat of Al7075 base material reinforced with SiC, Al2O3, B4C, TiO_2, ZnO, and SiO_2 (Hentschke, 2016). The same trend is observed for all the nanocomposites (Al7075-ZnO, Al7075-TiO_2, Al7075-Al_2O_3, and Al7075-SiC) as that observed for Al6061 nanocomposites.

Figure 22. Specific heat variations of Al6061-Al_2O_3, Al6061-B_4C, Al6061-SiC, Al6061-SiO_2, Al6061-TiO_2, and Al6061-ZnO

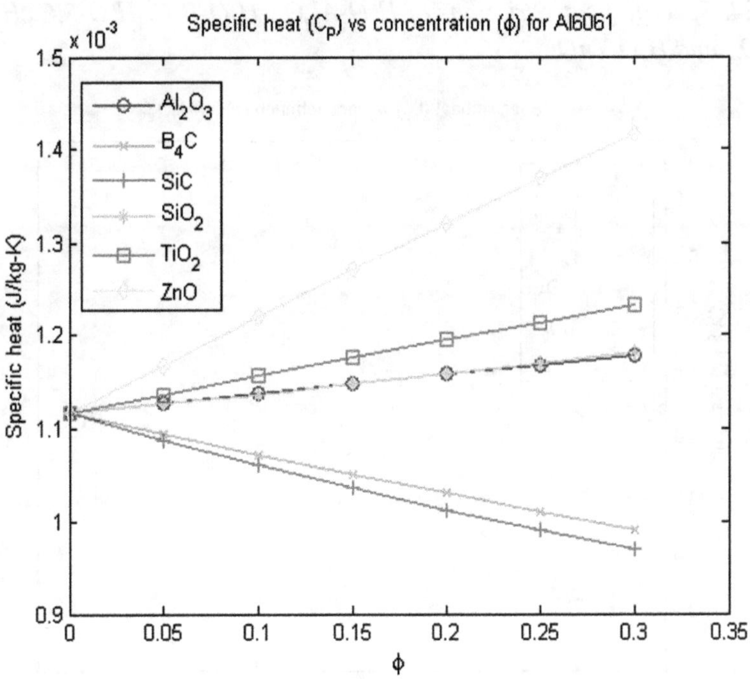

Figure 23. Specific heat variations of Al7075-Al_2O_3, Al7075-B_4C, Al7075-SiC, Al7075-SiO_2, Al7075-TiO_2, and Al7075-ZnO

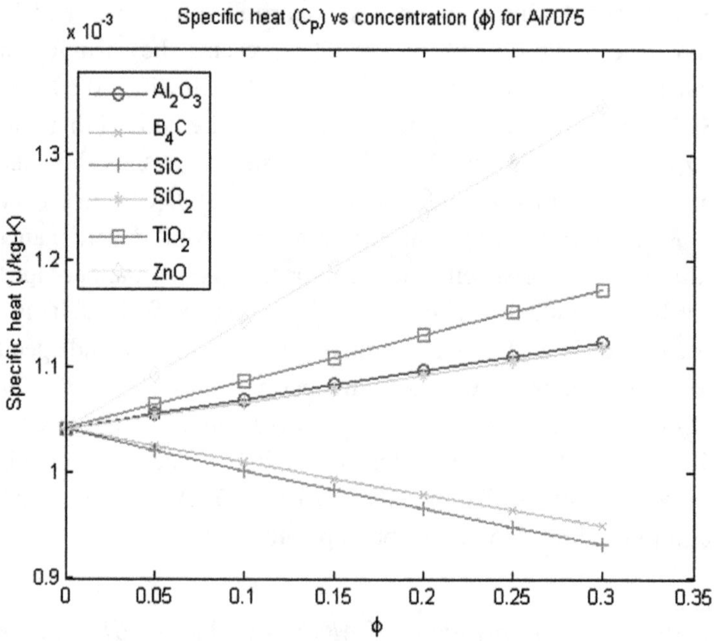

Figure 24. Specific heat variations of H_2O-Al_2O_3, H_2O-B_4C, H_2O-SiC, H_2O-SiO_2, H_2O-TiO_2, and H_2O-ZnO

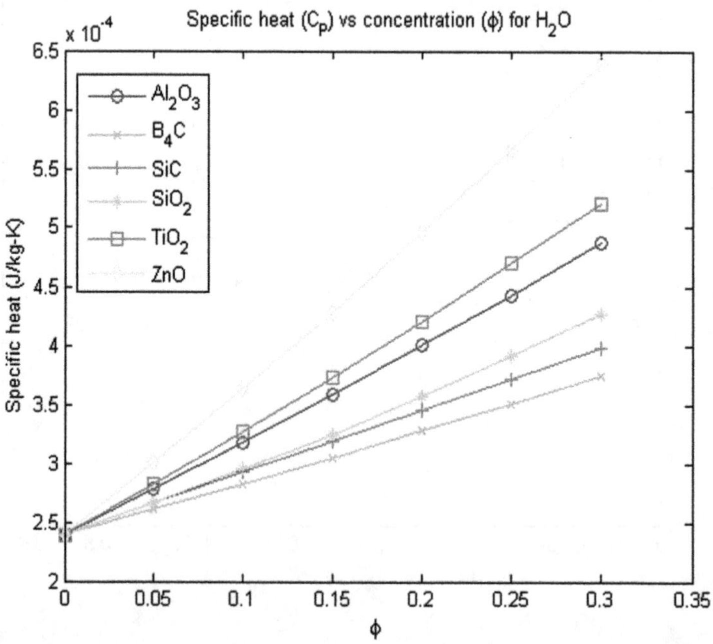

Figure 24 represents the variation of effective specific heat of H_2O base material reinforced with SiC, Al2O3, B4C, TiO_2, ZnO, and SiO_2. From the figure, it can be noticed that all the nanocomposites show increment in the specific heat. This may be due to the fact that effective nanocomposite of specific heat and density of all the nanocomposites is higher than the nanocomposite of specific heat and density of base material H_2O. From the figure, it can be noticed H_2O-ZnO nanocomposite shows maximum effective density compared to other nanocomposites while that of H_2O-B_4C nanocomposite shows minimum density (Hentschke, 2016). This can be due to the fact that specific heat and density of ZnO nanoparticle is highest while specific heat and density of SiC nanoparticle are lowest when compared with other nanoparticles.

CONCLUSION

Development of miniature electronic devices with the advancement in technology is in the requirement of advanced materials which should be lighter in weight and have higher energy transfer capability. Nanocomposites are the new era material that can fulfill these requirements. Present work deals with the numerical study of these nanocomposites. Six different nanoparticles studied in the present work are SiC, Al2O3, B4C, TiO_2, ZnO, and SiO_2 while three base materials considered are Al6061, Al7075 and H_2O. These nanoparticles are reinforced into the base material with different concentration percentage. Reinforcement effect of these nanoparticles into the base material is analyzed by studying thermal conductivity, density, and specific heat.

- Nanocomposites higher effective thermal conductivity is suitable for high heat releasing energy devices.
- Addition of nanoparticle increases the surface are to volume ratio which further increases the energy transfer rate.
- Nanocomposites with lower effective density are suitable when there is a requirement of reduction in weight for same heat release applications.
- Increment in the nanoparticle concentration further enhances the nanocomposites effective properties.
- Nanolayer thickness also affects the effective thermal conductivity.
- Nanoparticles shape factor also affect the effective thermal conductivity.

REFERENCES

Abu-Nada, E., & Oztop, H. F. (2009). Effects of Inclination Angle on Natural Convection in Enclosures Filled With Cu–Water Nano-Fluid. *International Journal of Heat and Fluid Flow*, *30*(4), 669–678. doi:10.1016/j.ijheatfluidflow.2009.02.001

Al-Shamani, A. N., Yazdi, M. H., Alghoul, M. A., Abed, A. M., Ruslan, M. H., Mat, S., & Sopian, K. (2014). Nanofluids for Improved Efficiency in Cooling Solar Collectors: A Review. *Renewable & Sustainable Energy Reviews*, *38*, 348–367. doi:10.1016/j.rser.2014.05.041

Bernard, O., Ulm, F. J., & Lemarchand, E. (2003). A Multiscale Micromechanics-Hydration Model for the Early-Age Elastic Properties of Cement-Based Materials. *Cement and Concrete Research*, *33*(9), 1293–1309. doi:10.1016/S0008-8846(03)00039-5

Bigley, C., & Greenwood, P. (2003). Using Silica to Control Bleed and Segregation in Self-Compacting Concrete. *Concrete (London)*, 43–45.

Bilgen, E., & Muftuoglu, A. (2008). Natural Convection in an Open Square Cavity with Slots. *International Communications in Heat and Mass Transfer*, *35*(8), 896–900. doi:10.1016/j.icheatmasstransfer.2008.05.001

Brinkman, H. C. (1952). The Viscosity of Concentrated Suspensions and Solutions. *The Journal of Chemical Physics*, *20*(4), 571. doi:10.1063/1.1700493

Cientifica, C. M. P. (2002). *The Nanotechnology Opportunity Report*. Retrieved from www.cmp-cientifica.com/

Einstein, A. (1906). Eine Neue bestimmung der molekuldimensionen. *Annalen Der Physik, Leipzig*, *19*(2), 289–306. doi:10.1002/andp.19063240204

Einstein, A. (1956). *Investigation on Theory of Brownian Motion*. Dover.

Garboczi, E. J., & Bentz, D. P. (1997). Analytical formulas for Interfacial Transition Zone Properties. *Journal of Advanced Cement Based Materials*, *6*(3-4), 99–108. doi:10.1016/S1065-7355(97)90016-X

Hamilton, R. L., & Crosser, O. K. (1962). Thermal Conductivity of Heterogeneous Two-Component System. *Industrial & Engineering Chemistry Fundamentals*, *1*(3), 187–191. doi:10.1021/i160003a005

Hentschke, R. (2016). On the specific heat capacity enhancement in nanofluids. *Nanoscale Research Letters*, *11*(1), 88. doi:10.118611671-015-1188-5 PMID:26873263

Kabir, E., Kumar, V., Kim, K. H., Yip, A. G. K., & Sohn, J. R. (2018). Environmental Impacts of Nanomaterials. *Journal of Environmental Management*, *225*, 261–271. doi:10.1016/j.jenvman.2018.07.087 PMID:30096714

Khanafer, K., Vafai, L., & Lightstone, M. (2003). Buoyancy-Driven Heat Transfer Enhancement in a Two-Dimensional Enclosure Utilizing Nano-Fluids. *International Journal of Heat and Mass Transfer*, *46*(19), 3639–3653. doi:10.1016/S0017-9310(03)00156-X

Khanafer, K. M., Al-Amiri, A. M., & Pop, I. (2007). Numerical Simulation of Unsteady Mixed Convection in a Driven Cavity using an Externally Excited Sliding Lid. *European Journal of Mechanics - B/Fluids*, *26*(5), 669–687. doi:10.1016/j.euromechflu.2006.06.006

Kishi, T., & Maekawa, K. (1997). Multi-Component Model for Heat of Hydration Heating of Blended Cement with Blast Furnace Slag and Fly Ash. *Concrete Library of JSCE*, *30*(12), 125–139.

Kosti, S. (2014). Numerical Study of Heat Flux Boundary in Nanofluid-Filled Cavity. *Nanomaterials and Energy*, *3*(6), 193–205. doi:10.1680/nme.14.00017

Kosti, S. (2019). Nanoparticle Reinforcement Effect on the MMC's. Advances in Interdisciplinary Engineering, 459-469.

Kosti, S., Das, M. K., & Saha, A. K. (2013). Buoyancy-Driven Flow and Heat Transfer in a Nano-Fluid-Filled Enclosure. *Nanomaterials and Energy*, *2*(4), 200–211. doi:10.1680/nme.13.00013

Kosti, S., & Malvi, C. S. (2018). Cumulative Influence of Nanoparticles on MMCs Time-Temperature History Curve. *Nanomaterials and Energy*, *7*(1), 1–10. doi:10.1680/jnaen.17.00011

Kosti, S., & Pathak, P. (2018). Genetic Algorithm Based Finite Difference Simulation of Solidification Process of MMC's. *Materials Today: Proceedings*, *5*, 8271–8279.

Kuzumaki, T., Miyazawa, K., Ichinose, H., & Ito, K. (1998). Processing of Carbon Nanotube Reinforced Aluminium Composites'. *Journal of Materials Research*, *13*(9), 2445–2449. doi:10.1557/JMR.1998.0340

Lai, F. H., & Yang, Y. T. (2011). Lattice Boltzmann Simulation of Natural Convection Heat Transfer of Al2O3/Water Nano-Fluids in a Square Enclosure. *International Journal of Thermal Sciences*, *50*(10), 1930–1941. doi:10.1016/j.ijthermalsci.2011.04.015

Lee, J.-H., Lee, S.-H., Choi, C. J., Jang, S. P., & Choi, S. U. S. (2010). A Review of Thermal Conductivity Data, Mechanism and Models for Nanofluids. *International Journal of Micro-Nano Scale Transport*, *1*(4), 269–322. doi:10.1260/1759-3093.1.4.269

Masciangioli, T., & Zhang, W. X. (2003). Environment Technologies at the Nano-scale. *Environmental Science & Technology*, *37*(5), 102A–109A. doi:10.1021/es0323998 PMID:12666906

Maxwell, J. C. (1873). *A Treatise on Electricity and Magnetism* (Vol. 1). Clarendoon Press.

Moustafa, A. F. (2017). Isothermal Reduction Process and Kinetic of Nanomaterials in Reducing Atmosphere: A review. *Journal of Analytical and Applied Pyrolysis*, *127*, 126–139. doi:10.1016/j.jaap.2017.08.015

Rohatagi, P. K. (1993). Metal Matrix Composites. *Defence Science Journal*, *43*(4), 323–349. doi:10.14429/dsj.43.4336

Rubel, A., & Landis, F. (1969). Numerical Study of Natural Convection in a Vertical Rectangular Enclosure. *Physics of Fluids*, *12*, 208–213.

Santosh, C., Velmurugan, V., Jacob, G., Jeong, S. K., Grace, A. N., & Bhatnagar, A. (2016). Role of Nanomaterials in Water Treatment Applications: A Review. *Chemical Engineering Journal*, *306*, 1116–1137. doi:10.1016/j.cej.2016.08.053

Santra, A. K., Sen, S., & Chakraborty, N. (2008). Study of Heat Transfer Augmentation in a Differentially Heated Square Cavity using Copper–Water Nano-Fluid. *International Journal of Thermal Sciences*, *47*(9), 1113–1122. doi:10.1016/j.ijthermalsci.2007.10.005

Sarvia, R. M., & Fuskele, V. (2017). Review on Thermal Conductivity of Nanofluids. *Materials Today: Proceedings*, *4*(2), 4022–4031.

Schwartz, L., Garboczi, E., & Bentz, D. (1995). Interfacial Transport in Porous Media: Applications to DC Electrical Conductivity of Mortars. *Journal of Applied Physics*, *78*(10), 5898–5908. doi:10.1063/1.360591

Sharma, V. P., Sharma, U., Chattopadhyay, M., & Shukla, V. N. (2018). Advance Applications of Nanomaterials: A review. *Materials Today: Proceedings*, *5*, 6376–6380.

Singh, M., Lara, S., & Tlali, S. (2017). Effects of Size and Shape on the Specific Heat, Melting Entropy and Enthalpy of Nanomaterials. *Journal of Taibah University for Science Journal*, *11*(6), 922–929. doi:10.1016/j.jtusci.2016.09.011

Talebi, F., Mahmoudi, A. H., & Shahi, M. (2010). Numerical Study of Mixed Convection Flows in a Square Lid-Driven Cavity Utilizing Nano-Fluid. *International Communications in Heat and Mass Transfer*, *37*(1), 79–90. doi:10.1016/j.icheatmasstransfer.2009.08.013

Tay, C. Y., Setyawati, M. I., & Leong, D. T. (2017). Nanoparticle Density: A Critical Biophysical Regulator of Endothelial Permeability. *ACS Nano*, *11*(3), 2764–2772. doi:10.1021/acsnano.6b07806 PMID:28287706

Timofeeva, E. T., Routbort, J. L., & Singh, D. (2009). Particle shape effect on thermophysical properties of alumina nanofluids. *Journal of Applied Physics*, *106*(1), 1–10. doi:10.1063/1.3155999

Tiwari, R. K., & Das, M. K. (2007). Heat Transfer Augmentation in a Two-Sided Lid-Driven Differentially Heated Square Cavity Utilizing Nano-Fluids. *International Journal of Heat and Mass Transfer*, *50*(9-10), 2002–2018. doi:10.1016/j.ijheatmasstransfer.2006.09.034

Turan, O., Poole, R. J., & Chakraborty, N. (2012). Influences of Boundary Conditions on Laminar Natural Convection in Rectangular Enclosures with Differentially Heated Side Walls. *International Journal of Heat and Fluid Flow*, *33*(1), 131–146. doi:10.1016/j.ijheatfluidflow.2011.10.009

Xiao, H., & Liu, S. (2018). 2D Nanomaterials a Lubricant Additive: A Review. *Materials & Design*, *135*, 319–332. doi:10.1016/j.matdes.2017.09.029

Yu, W., & Choi, S. U. S. (2003). The Role of Interfacial Layers in the Enhanced Thermal Conductivity of Nanofluids: Arenovated Maxwell Model. *Journal of Nanoparticle Research*, *6*(4), 167–171. doi:10.1023/A:1024438603801

Zhang, G. D. (2003). Electrical Properties of Nano-ceramics Reinforced with Ropes of Single-Walled Carbon Nanotubes, *Applied Physics Letters, 83,* 1228-1230. doi:10.1063/1.1600511

Chapter 2
Mechanical Behaviour of Carbon Nanotubes

Danilo Vuono
Federiciana Università Popolare, Italy

ABSTRACT

Mechanical properties of carbon nanotubes (CNTs) are very interesting for the nanocomposite field. The possibility to use these nanomaterials as fibers in polymeric matrix is one of the most important applications of the last years. This study recognised the mechanical properties of CNTs and the obtained composites with polymeric matrix. The second part of chapter presents a short characterisation of an epoxy-CNTs-based composite. This study verified the hypothesis made concerning a different tensile strength of these materials as a function of presence of structure defects.

INTRODUCTION

Carbon nanotubes are molecular carbon fibers, formed by little graphitic cylinders closed to the extremity by caps of fullerenes. Carbon nanotubes are classified in two families: multi-walled nanotubes (MWCNTs), discovered in 1993 (Ijima & Ichihashi, 1993) and single-walled nanotubes (SWCNTs), observed for the first time by Bethune et al. (Bethune et al. 1993). SWCNTs can be imagined as a graphitic plane rolled upon itself, formed by hexagonal sweaters of carbon atoms with sp^2 hybridisation (Iijima, 1991). CNTs have several structural configurations: armchair, zig zag and chiral. Different configurations have different physico-chemical properties, especially for electric conductivity. MWCNTs are formed by several concentric graphitic sheets. CNTs are very flexible and hard materials. The secret of these

DOI: 10.4018/978-1-7998-1530-3.ch002

properties is in the σ carbon-carbon bonds that are present in the graphite (Iijima, 1991; Chiang et al. 2001). Hence, their applications as fibers in nanocomposite materials is developed in the last years (Chiang et al. 2001). Moreover, they are considered as very good thermal conductors with high performances comparable to those of graphite (Popov, 2004). They are also applied in the field of electrical conductors. The SWCNTs can assume a metallic or semiconductor behaviour as a function of the rolling up of the graphitic plane to create the tube (Ajayan & Ebbesen, 1997). In so far, the zig-zag CNTs could always considered as semiconductors and sometimes as metallic conductors in particular structural conditions, while the armchair CNTs are always considered as metallic conductors (Frank et al. 1998). CNTs have presented high sensibility to the electric fields, folding up itself up to 90° and then taking back their initial shape without damages (Baughman & Zakhidov & de Heer, 2002). Carbon nanotubes have a high chemical inactivity. They withstand to the oxidative agents and acids. They present a strong capillary property. Therefore SWCNTs, for instance, are used as ideal adsorbent materials for liquids and gases (Pederson & Broughton, 1992). Adsorption properties of CNTs have been studied for the H_2 adsorption in order to optimise new storage techniques (Piérard et al. 2002). Adsorption is a superficial molecular attraction phenomenon between the solid phase, the adsorbent, and a gas or liquid phase, the adsorbate. Adsorption process in a solvent-solute-solid system is strongly affected both by solvent-solute affinity and solute-adsorbent interactions. This is realised through three types of forces: electrostatic forces, van der Waals forces and chemical forces. Li et al. studied Pb, Cd and 1,2-dichlorobenzene adsorption using carbon nanotubes. Indeed, results demonstrate that nanotubes have remarkable adsorption capacities (Li et al. 2007). In particular, the CNTs with structural defects have high adsorption performances, compared to those without structural defects. The interest for this kind of application has allowed to develop several computer simulations using analytical theoretical models (Furmaniak et al. 2006). The interaction between the CNTs and the organic compounds are strongly dependent on nanotubes diameters and, of course, the adsorbent-adsorbate interactions. Liao et al. studied the adsorption properties of the CNTs for phenolic derivates (Liao & Sun & Gao, 2007). Yang et al. detected the adsorption capacities of MWCNTs for the APH (Aromatic Polycyclic Hydrocarbons) (Yang & Zhu & Xing, 2006; Yang & Xing 2006; Yang et al. 2006). In 2009, a Russian research group tested the adsorption interactions between multi-walled carbon nanotubes and benzoic acid in aqueous solutions (Kotel et al. 2009). They determined the optimal functionalised CNTs to be used in this application. Nevertheless, applications using their mechanical properties remain the main field of development for these kinds of materials. Since 1995, Ruoff and Lorents (1995) discussed aspects of the mechanical and thermal properties of carbon nanotubes. The tensile and bending stiffness constants of ideal multi-walled and single-walled carbon nanotubes are derived in terms of the known elastic properties of graphite. Tensile strengths are estimated by scaling the 20 GPa tensile strength of Bacon's graphite whiskers. It is widely perceived that carbon nanotubes will allow construction of composites with extraordinary strength: weight ratios, due to the inherent strength of the nanotubes. Several "rules of thumb" have been developed in the study of fiber/matrix composites. Close inspection of these shows that carbon nanotubes satisfy several criteria, but that others remain untested (Ruoff & Lorents, 1995). Campbel et al., in 1999, report here the fabrication and characterization of electrodes constructed from single carbon nanotubes. The sigmoidal voltammetric response of these nanotubular electrodes is characteristic of steady-state radial diffusion. They demonstrated that electrochemical nanotubular electrodes can be constructed from single carbon nanotubes. Insulated electrodes of arbitrary length with 80-200 nm diameters can be routinely fabricated. These electrodes represent a new application of carbon nanotubes that takes advantage of their geometrical shape, mechanical strength, and electrical conductivity (Campbell & Sun & Crooks, 1999). Finally, Salvetat et al. present variety of outstanding experimental results on the elucidation of the elastic properties of carbon nanotubes are fast appearing. These are based mainly on the techniques of high-resolution transmission electron microscopy (HRTEM) and atomic force microscopy (AFM) to determine the Young's moduli of single-wall nanotube bundles and multi-walled nanotubes, prepared by a number of methods. Techniques are used for these objectives for the first time. Collected data show that the Young's modulus of CNTs is at least as high as graphite and can be even higher for small SWNTs. Experiments show that Young's moduli for MWNTs are dependent upon the degree of order within the tube walls.

As a conclusion, it is possible to note schematic representation of these findings, where the Young's modulus decreases as the disorder increases (Salvetat et al. 1999). In the last years, the research on these potential applications of CNTs are increased and papers concerning these topics are more and more.

Mechanical Strength of CNTS

SWCNTs are predicted to have extremely high Young's modulus values, similar to that of graphite in-plane (~1000 GPa) (Kelly, 1981). It is also predicted that SWCNTs can sustain large strain in the axial direction, and preferred strain-releasing mechanisms have been recently analyzed (Yakobsen, 1997; Yakobson 1998; Nardelli et al. 1998). High breaking strengths are thus predicted for SWCNTs, generating interest for high strength, lightweight material applications. A recent model shows that a strained SWCNT can readily release strain energy by defect nucleation beyond a critical tensile strain of about 5%. In contrast to the extensive theoretical modeling, there have been few experimental studies on the mechanical properties of SWCNTs. Some research produced the method used to measure the strength of individual multiwalled carbon nanotubes (MWCNT) to SWCNT ropes. SWCNT ropes were tensile loaded in a "nanostressing stage" operated inside a scanning electron microscope (SEM). Min-Feng Yu et al. proposed two separate treatments for calculating the cross sectional area to be used for determining the applied stress give low-end and high-end values for the breaking strength and Young's modulus values of these SWCNT ropes (Fig.1). In the first treatment, every SWCNT in the rope is assumed to carry an equal load. Then, the cross-sectional area is defined as the total number of SWCNTs in the rope multiplied by the cross-sectional area of a SWCNT. In the second treatment, only the perimeter SWCNTs in the close-packed SWCNT rope are assumed to carry the initial load, so the load-bearing cross-sectional area is equal to the total number of SWCNTs on the perimeter of the rope multiplied by the cross-sectional area of a SWCNT (Yu et al. 2000).

Segregation Issues of CNTs in Polymeric Nanocomposites

Mamedov et al. Studied the exceptional mechanical properties of single-wall carbon nanotubes (SWNT) have prompted intensive studies of SWNT composites. However, the present composites have shown only a moderate strength enhancement when compared with other hybrid materials. Although substantial advances have been made, mechanical properties of SWNT-doped polymers are noticeably below their

Figure 1. Eight stress versus strain curves obtained from the tensile-loading experiments on individual SWCNT ropes
(Yu et al. 2000)

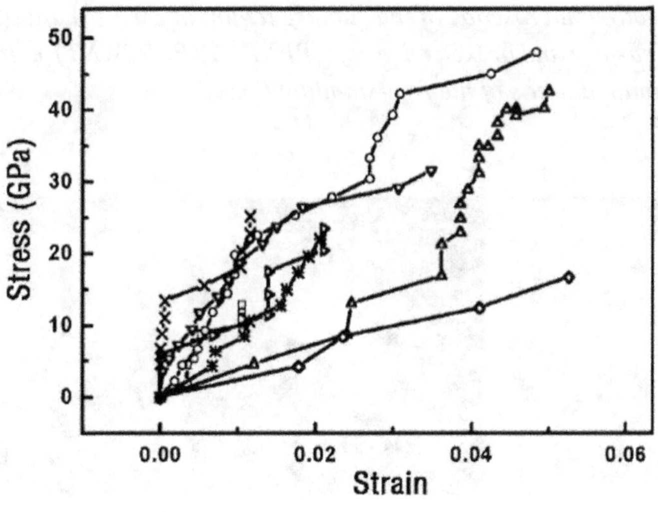

highly anticipated potential. Pristine SWNTs are well known for poor solubilization, which leads to phase segregation of composites. Severe structural inhomogeneities result in the premature failure of the hybrid SWNT/polymer materials (Fig.2). The mechanical properties of the layered composites were tested on a custom-made thin-film tensile strength tester (McAllister) recording the displacement and applied force by using pieces cut from $((PEI/PAA)(PEI/SWNT)_5)_6$ and $((PEI/PAA)(PEI/SWNT)_5)_8$ freestanding films (Mamedov et al. 2002).

Segregation problems can be successfully mitigated when the SWNT composite is made following the protocol of layer-by-layer assembly. This deposition technique prevents phase segregation of the polymer/SWNT binary system, and after subsequent crosslinking, the nanometre-scale uniform composite with SWNT loading as high as 50 wt% can be obtained. The freestanding SWNT/polyelectrolyte membranes delaminated from the substrate were found to be exceptionally strong with a tensile strength approaching that of hard ceramics. Because of the lightweight nature of SWNT composites, the prepared free-standing membranes can serve as components for a variety of long-lifetime devices (Mamedov et al. 2002).

Mechanical Properties of CNTS Effected by Structural Defects

As-grown carbon nanotubes have relatively few defects. However, defects can appear at the purification stage or be deliberately introduced by irradiation with energetic particles or by chemical treatment when aiming at the desired functionality. The defects, especially vacancies, give also rise to an effect of deterioration of axial mechanical properties of nanotubes. By employing molecular dynamics simulations and continuum theory, we study how the Young's modulus and tensile strength of

Figure 2. Electron microscopy of the rupture region in SWNT multilayers. a, SEM image of the surface and broken edges of $((PEI/PAA)(PEI/SWNT)_5)_8$. b, and c, TEM images of ruptured areas of the free-standing films
(Mamedov et al. 2002)

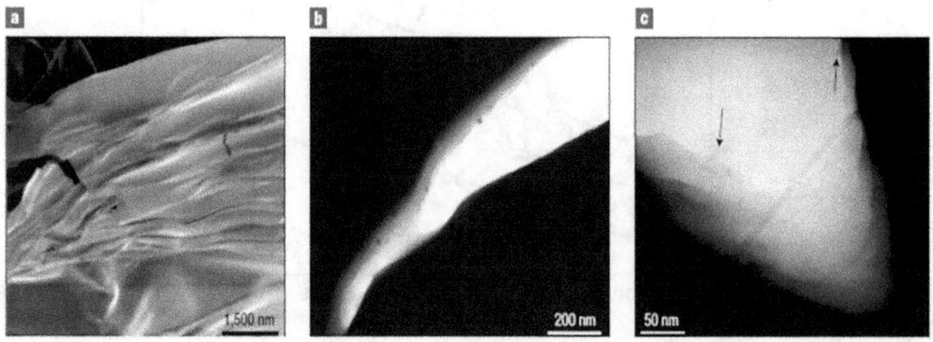

nanotubes with vacancy-related defects depend on the concentration of defects and defect characteristics (Fig.3).

Final results indicate that the Young's modulus of nanotubes with defects will essentially be the same unless the vacancy concentration is extremely high. On the other hand, the tensile strength will substantially drop due to the quasi-one-dimensional atomic structure of SWNTs already if a single vacancy is present the tensile strength of a SWNT is governed by the "weakest" segment of the tube. Given that a small number of defects are always present in nanotubes, this may explain why the theoretically predicted Young's modulus agrees well with the experimentally measured values, while the tensile characteristics are much worse (Sammalkorpi et al. 2004).

Figure 3. Different kinds of structural defects of SWCNTs. nonreconstructed (a)–(c) and reconstructed (d)–(i) single (a),(d),(g), double (b),(e),(h) and triple (c),(f),(i) vacancies
(Sammalkorpi et al. 2004)

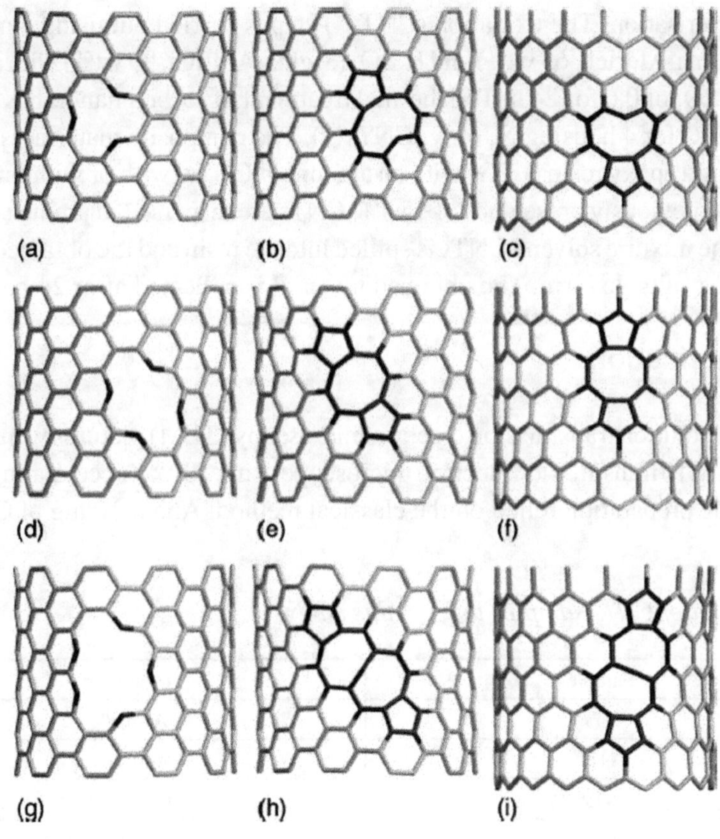

EXPERIMENTAL

Preparation of Samples

The supported catalyst used for the production of carbon nanotubes is a Co-Fe on NaY (UOP Y-54 DR PWD) zeolite (Hernadi et al. 1996). The metallic percentage is of 5 wt % for each. The catalysts were prepared by impregnation method on the powdered zeolitic phase. The synthesis method used for the massive production of $CNTs^{NP}$ (no purified carbon nanotubes) is the Catalytic Chemical Vapour Deposition (CCVD) method. The source of carbon is ethylene using N_2 as carrier gas. The synthesis temperature is 700 °C with a reaction time of 20 min. The carbon deposit (Cd) of this reaction is of 1447%. The nanotubes purification ($CNTs^{P}$) is performed by dissolving the zeolitic support in HF (Sigma-Aldrich, 40 wt%) to avoid the presence of metallic oxides with several steps of washing. 1 g of $CNTs^{NP}$ is added to 200 ml of HF and omogenised. This treatment takes 3 days. Carbon nanotubes are filtered and treated with other 200 ml of HF for 24 h. The so-obtained product is washed by distilled water and dried in oven for ca.18 h. $CNTs^{P}$ are collected for the characterisation. The oxidation ($CNTs^{ox}$) step is carried out using a mixture of HNO_3 (Sigma-Aldrich, 65 wt%) and H_2SO_4 (Sigma-Aldrich, 99 wt%) with a v/v ratio HNO_3/ H_2SO_4 of 0.6 for 24 h. The thermal treatment of carbon nanotubes is carried out at 900 °C for 4 h, using N_2 only ($CNTs^{900}$). The composite material is prepared using a liquid epoxy resin (EPO) with an amount of CNTs (with or without different treatments previously presented) of 1 wt%. CNTs are dispersed in amino solvent for 10 min. The mixture solvent-CNTs is spilled into the resin and the obtained material is dried in a suitable form. The obtained material is collected after 24 hours.

Characterisation

The low-resolution transmission electron microscopy (TEM) pictures were taken on a Philips CM10 transmission electron microscope using 100 kV accelerating voltage. The sample preparation relied on the classical method. About 10 mg of CNT were

Table 1. List of CNTs samples used in this study.

Sample	Treatment
$CNTs^{NP}$	As-made
$CNTs^{P}$	Purification
$CNTs^{OX}$	Oxidation
$CNTs^{900}$	Thermal treatment

suspended in 3 ml ethanol and the suspension was then deposited on a carbonated Cu-Rh grid. Several pictures were analysed using the Soft Imaging Viewer software. The morphology of the products was examined on a scanning electron microscope (FEI model Inspect). The thermo-analytical measurements were performed on the automatic TG/DTA instrument under air flow (50 cc/min with heating rate of 10 °C min-1). The tensile tests are carried out using sample of 10 cm and a 0.3 cm of width. The ends of samples are larger than the rest of piece.

Discussion

Obtained product from the decomposition of ethylene by CCVD method was characterised to determine its initial properties before the following treatment processes. Figure 4 shows thermal analysis data of the as-synthesised carbonaceous material. Tg curve points out a weight loss of ca. 94%, while the presence of a unique peak in the DTg curve at ca. 610 °C demonstrates that the total weight loss is due to a single thermal effect (Fig. 4-a). Figure 4-b presents the DTA curve of the sample. An exothermal peak can be noted at 605 °C, clearly connected to the weight loss. This behaviour suggests that the material is not completely burned. In fact, the total weight loss of the analysed CNTs is not equal to 100%. This means that the catalyst is still present in the sample, with a composition of ca. 6 wt % of the total weight. The value of exothermal peak of DTA curve, due to the combustion of carbon nanotubes, is indicative of a good graphitisation of the product, while the lack of exothermal peaks at lower temperatures than 600 °C denotes the absence of amorphous phases (in the range 300-350 °C) or synthesised products with low graphitisation quality (range 350-450 °C). Moreover, the sharp shape of the DTA peak highlights a homogeneous graphitisation of the sample.

Purification treatment, using the method described in experimental section, is suitable to free the graphitised product from the presence of the supported catalyst. The use of a HF based process is useful “to wash” the siliceous phase of support and disperse the metallic nanoparticles facilitating the washing by distilled water. Figure 5 presents the Tg and DTg curves (Fig. 5-a) and DTA curve (Fig. 5-b) of a purified sample of CNTs.

It is important to note the well-done purification step. Indeed, the total weight loss corresponds to 100 wt%. This means that the product is totally burned. The presence of two peaks in the DTg curve attests that the thermal phenomena is due to two different effects: the first one at ca. 605°C and the second one at ca. 620 °C. Two different relative weight losses of 80% and 20%, respectively, are connected to these thermal effects. The DTA curve shows two exothermal peaks at the same temperatures. The shape of the first one is very sharp and this is due to a high definition of the graphitization properties of the sample after the purification. The

Figure 4. Tg, DTg (a) and DTA (b) curves of unpurified carbon nanotubes

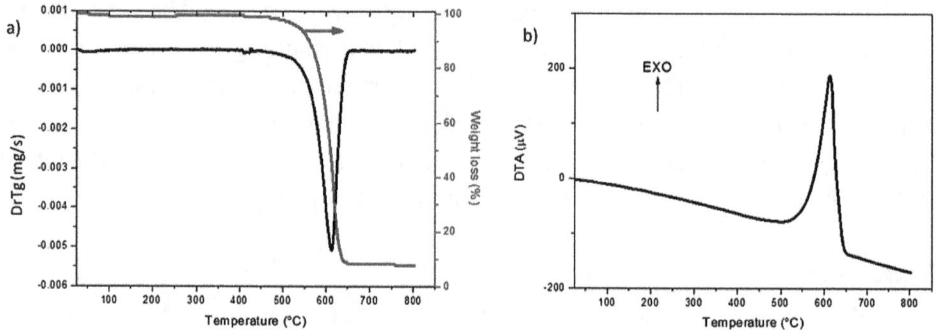

Figure 5. Tg, DTg (a) and DTA (b) curves of purified carbon nanotubes

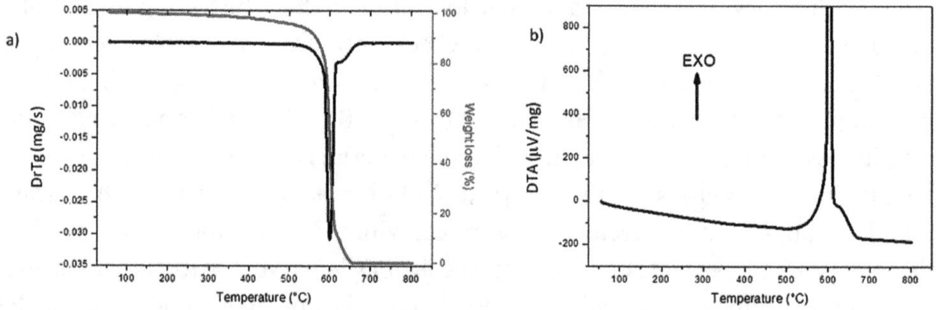

presence of an exothermal peak at higher temperature denotes the presence of 20 wt% of the sample with a very high graphitisation. Thus, it is possible to affirm that the purification process improves the graphitisation properties of the CNTs, somehow hidden by the presence of the catalyst, as impurity. Figure 6 shows the morphology of the obtained product. In confirmation of the thermal analysis results, no any trace of amorphous carbon is present.

The morphology of CNTs are shown as filaments of ca. 20 µm of length and ca. 20 nm of external diameter. The number of concentric walls is ca. 5-7.

The thermal analyses of the oxidised CNTs are shown in Figure 7. In this case, the total weight loss is 100%. This is due to different thermal effects, as confirmed by the four DTg peaks (Fig. 4-a). The first relative weight loss is connected to a DTg peak at ca. 100°C, the second DTg peak is at ca. 450 °C and the last two weight losses are connected to two DTg peaks at 600 °C and 690 °C, respectively. DTA curve points out an endothermal peak at 100 °C, determined by the water loss of the sample (Fig. 7-b). This phenomenon is not found in other kinds of samples of CNTs, classically hydrophobic, and it could be connected to a certain affinity with H_2O molecules of

Figure 6. TEM image of purified sample of CNTs

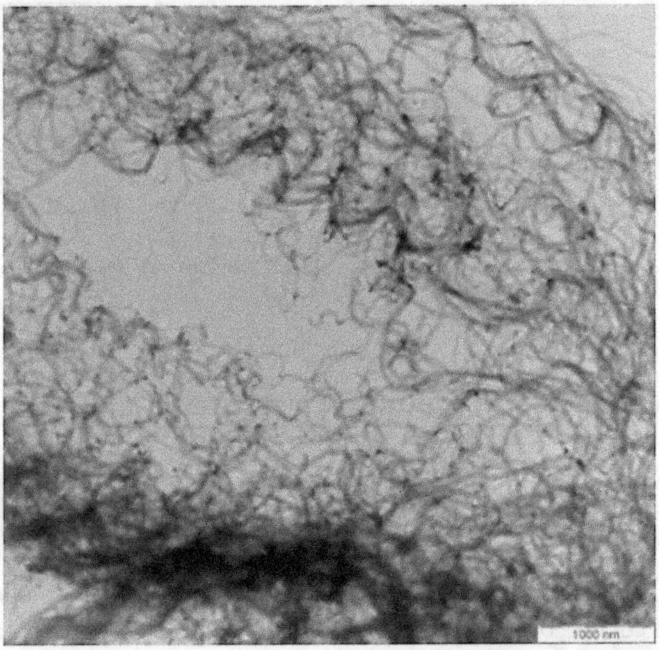

the oxidised CNTs. The DTA peak at 450°C, exothermal, highlights the presence of part of the sample at low graphitisation level. This behaviour, remembering the trends of the DTA curves of not purified and purified CNTs discussed above, could be due to the introduction of a high number of –OH, -COOH and –COR groups during the oxidisation process, decreasing the thermal stability of the CNTs. In fact, the initial quality of the samples denotes a high degree of graphitisation, while the oxidation treatment reduces this property. The weight loss connected to the desorption of water is 5%. The second weight loss reaches 15%. The last two DTA peaks are connected to an optimal graphitisation level. Nevertheless, the shape of the DTA peaks, exothermal, at 600 °C and 650 °C, respectively, seems to be not very sharp. Moreover, the second peak is presented as a shoulder of the first one. This could be due to a distribution of the graphitisation level of the sample related to the oxidisation treatment and not to a homogeneous graphitisation like in the not purified CNTs, for instance.

Figure 7. Tg, DTg (a) and DTA (b) curves of oxidised carbon nanotubes

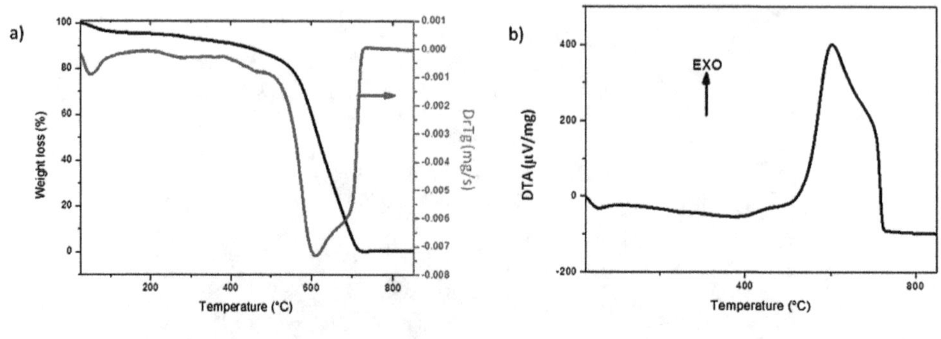

Use of MWCNTs as Nanofibers in Polymeric Matrix to Produce Nanocomposite Materials

Figure 8 shows the tensile behaviour of nanocomposites produced using different kinds of CNTs. The EPO sample represents the polymeric material without CNTs as nanofibers. The percentage of several kinds of used CNTs is maintained at 1 wt%. This is due to define better the mechanical differences of samples as a function of CNTs produced by different treatments. The sample with 1 wt% of as-made CNTs presents the best fitting. It is possible to note a trend to increase the resistance

Figure 8. Tensile strength of nanocomposites produced using different CNTs samples as fibers

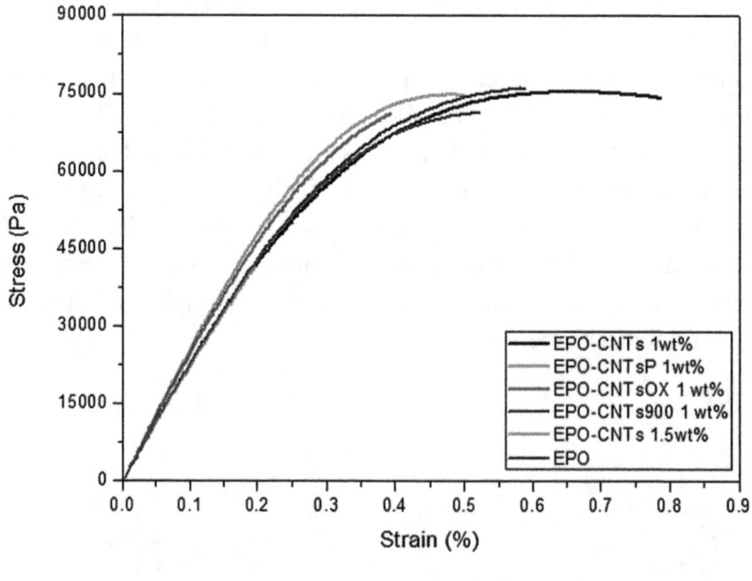

properties but, maintaining the σ_{max} at ca. 75000 Pa. This behaviour is decreasing in the samples using purified, oxidised and CNTs treated at 900°C, respectively. This could be due to an increase of structural defects, induced by several and different treatments carried out on the CNTs used as nanofibers. In particular, the CNTs treated at 900°C appears with no different mechanical behaviours than those of EPO with CNTs as-made. The hypothesis could be that CNTs used in this application after the purification and oxidation step could present a certain number of defects. These defects could be destroyed during the thermal treatment, leaving graphitised structure similar to pristine CNTs and determining mechanical properties similar to those of polymeric matrix with as-made CNTs.

CONCLUSION

The study presents a short characterisation of materials using several samples of MWCNTs after purification, oxidation, thermal treatment at 900 °C, respectively. The tensile properties of the obtained materials are tested and the best sample is defined. Sample with 1 wt% of as-made CNTs seems to have a better fitting of tensile strength compared to other nanocomposites produced with purified, oxidised ones. This could be due to the creation of structural defects after all of these treatments. This could be confirmed observing the tensile strength fit of material formed by epoxy resin and CNTs treated at 900 °C. In fact, thermal treatment deletes the structural part with defects and leaves the graphitised structure only. Thus, the tensile strength of this sample is similar to that of composite formed by epoxy resin and as-made CNTs. The possibility to produce defects could be important in other potential applications of CNTs, like adsorption process, but it does not determine good properties in nanotubes used as fibers in polymeric matrix, as previous studies confirm (Sammalkorpi et al. 2004).

REFERENCES

Ajayan, P. M., & Ebbesen, T. W. (1997). Nanometre-size tubes of carbon. *Reports on Progress in Physics*, *60*(10), 1025–1062. doi:10.1088/0034-4885/60/10/001

Baughman, R. H., Zakhidov, A. A., & de Heer, W. A. (2002). Carbon Nanotubes--the Route Toward Applications. *Science*, *297*(5582), 787–792. doi:10.1126cience.1060928 PMID:12161643

Bethune, D. S., Kiang, C. H., de Vries, M. S., Gorman, G., Savoy, R., Velazquez, J., & Beyers, R. (1993). Cobalt-catalysed growth of carbon nanotubes with single-atomic-layer Walls. *Nature*, *363*(6430), 605–607. doi:10.1038/363605a0

Campbell, J. K., Li, S., & Crooks, R. M. (1999). Electrochemistry Using Single Carbon Nanotubes. *Journal of the American Chemical Society*, *121*(15), 3779–3780. doi:10.1021/ja990001v

Chiang, I. W., Brinson, B. E., Huang, A. Y., Willis, P. A., Bronikowski, M. J., Margrave, J. L., ... Hauge, R. H. (2001). Purification and Characterization of Single-Wall Carbon Nanotubes. *The Journal of Physical Chemistry B*, *105*(35), 8297–8301. doi:10.1021/jp0114891

Frank, S., Poncharal, P., Wang, Z. L., & de Heer, W. A. (1998). Carbon nanotube quantum resistors. *Science*, *1744*, 280.

Furmaniak, S., Terzyk, A. P., Gauden, P. A., & Rychlicki, G. (2006). Simple models of adsorption in nanotubes. *Journal of Colloid and Interface Science*, *295*(2), 310–317. doi:10.1016/j.jcis.2005.12.032 PMID:16427068

Hernadi, K., Fonseca, A. B., Nagy, J., Bernaerts, D., Fudala, A., & Lucas, A. A. (1996). Catalytic synthesis of carbon nanotubes using zeolite support. *Zeolites, 17*, 416.

Iijima, S. (1991). Synthesis of Carbon Nanotubes. *Nature*, *354*, 56. doi:10.1038/354056a0

Iijima, S., & Ichihashi, T. (1993). Single-shell Carbon Nanotubes of 1-nm diameter. *Nature*, *363*(6430), 603–605. doi:10.1038/363603a0

Kelly, B. T. (1981). *Physics of Graphite*. London: Applied Science.

Kotel, L. Y., Brichka, A. V., & Brichka, S. Y. (2009). Adsorption properties of modified multilayer carbon nanotubes with respect to benzoic acid. *Russian Journal of Applied Chemistry*, *82*(4), 569–573. doi:10.1134/S1070427209040077

Li, Y. H., Zhao, Y. M., Hu, W. B., Ahmad, I., Zhu, Y. Q., Peng, X. J., & Luan, Z. K. (2007). Carbon nanotubes - the promising adsorbent in wastewater treatment. *Journal of Physics: Conference Series*, *61*, 698–702. doi:10.1088/1742-6596/61/1/140

Liao, Q., Sun, J., & Gao, L. (2007). The adsorption of resorcinol from water using multi-walled carbon nanotubes. *Colloids and Surfaces*, *312*(2-3), 160–165. doi:10.1016/j.colsurfa.2007.06.045

Mamedov, A. A., Kotov, N. A., Prato, M., Guldi, D. M., Wicksted, J. P., & Hirsch, A. (2002). Molecular design of strong single-wall carbon nanotube/polyelectrolyte multilayer composites. *Nature Materials*, *1*(3), 190–194. doi:10.1038/nmat747 PMID:12618809

Nardelli, M. B., Yakobson, B. I., & Bernholc, J. (1998b). Brittle and Ductile Behavior in Carbon Nanotubes. *Physical Review Letters*, *81*(21), 4780. doi:10.1103/PhysRevLett.81.4656

Nardelli, M. B., Yakobson, B. I., & Bernholc. (1998a). Mechanism of strain release in carbon nanotubes. *J. Phys. Rev. B*, *57*, R4277.

Pederson, M. R., & Broughton, J. Q. (1992). Nanocapillarity in fullerene tubules. *Physical Review Letters*, *69*(18), 2689–2692. doi:10.1103/PhysRevLett.69.2689 PMID:10046559

Popov, V. N. (2004). Carbon nanotubes: Properties and application. *Materials Science and Engineering*, *43*(3), 61–102. doi:10.1016/j.mser.2003.10.001

Ruoff, R. S., & Lorents, D. C. (1995). Mechanical And Thermal Properties Of Carbon Nanotubes. *Carbon*, *33*(7), 925–930. doi:10.1016/0008-6223(95)00021-5

Salvetat, J.-P., Bonard, J.-M., Thomson, N. H., Kulik, A. J., Forro, L., Benoit, W., & Zuppiroli, L. (1999). Mechanical properties of carbon nanotubes. *Applied Physics. A, Materials Science & Processing*, *69*(3), 255–260. doi:10.1007003390050999

Sammalkorpi, M., Krasheninnikov, A. V., Kuronen, A., Nordlund, K., & Kaski, K. (2005). Mechanical properties of carbon nanotubes with vacancies and related defects. *Physical Review B: Condensed Matter and Materials Physics*, *71*(16), 169906. doi:10.1103/PhysRevB.71.169906

Yakobson, B. I. (1997). Recent Advances in the Chemistry and Physics of Fullerences and Related Materials. Electrochemical Society.

Yakobson, B. I. (1998). Mechanical relaxation and "intramolecular plasticity" in carbon nanotubes. *Applied Physics Letters*, *72*(8), 918–920. doi:10.1063/1.120873

Yang, K., Wang, X., Zhu, L., & Xing, B. (2006). Competitive Sorption of Pyrene, Phenanthrene, and Naphthalene on Multiwalled Carbon Nanotubes. *Environmental Science & Technology*, *40*(18), 5804–5810. doi:10.1021/es061081n PMID:17007144

Yang, K., & Xing, B. (2006). Desorption of polycyclic aromatic hydrocarbons from carbon nanomaterials in water. *Environmental Pollution*, *145*(2), 529–537. doi:10.1016/j.envpol.2006.04.020 PMID:16777283

Yang, K., Zhu, L., & Xing, B. (2006). Adsorption of Polycyclic Aromatic Hydrocarbons by Carbon Nanomaterials. *Environmental Science & Technology*, *40*(6), 1855–1861. doi:10.1021/es052208w PMID:16570608

Yang, X., & Al-Duri, B. (2005). Kinetic modeling of liquid-phase adsorption of reactive dyes on activated carbon. *Journal of Colloid and Interface Science*, *287*(1), 25–34. doi:10.1016/j.jcis.2005.01.093 PMID:15914145

Yu, M. F., Files, B. S., Arepalli, S., & Ruoff, R. S. (2000). Tensile Loading of Ropes of Single Wall Carbon Nanotubes and their Mechanical Properties. *Physical Review Letters*, *84*(24), 5552–5555. doi:10.1103/PhysRevLett.84.5552 PMID:10990992

Chapter 3
Diverse Applications of Graphene-Based Polymer Nanocomposites

Pradip Majumdar
Northern Illinois University, USA

Amartya Chakrabarti
Dominican University, USA

ABSTRACT

Polymer nanocomposites are unique materials reinforced with nanoscale additives. Among a variety of nanomaterials available to act as filler additives in different polymer matrices, graphene is the most versatile one. Graphene-based polymer nanocomposites have improved electrical, mechanical, chemical, and thermal properties, which make them suitable for applications in the electronics, energy, sensor, and space sectors. Graphene, the nanosized filler, can be prepared using either a top-down or a bottom-up approach and dispersed in the polymer matrix utilizing different conventional techniques. The nanocomposite materials find usage in suitable area of applications depending on their specific characteristics. This chapter discusses the current state-of-the-art manufacturing techniques for graphene and graphene-based nanocomposite materials. Application of graphene-based polymer nanocomposites in the various fields with an emphasis on the areas high heat flux applications requiring enhanced thermal conductivity will be an additional major focus of this chapter.

DOI: 10.4018/978-1-7998-1530-3.ch003

INTRODUCTION

Since its discovery in the early twenty-first century (Novoselov 2004), graphene has been dominating the research and development sector of the materials world. The worldwide interest in graphene is clearly exhibited by the rapidly increasing number of publications in different format of available literatures (Figure 1). A Web of Science database topic search on graphene resulted in approximately 34000 published documents only in 2018, compared to the same of 2342 in 2009. This rapid growth in interest within last ten years can easily be attributed to the unique properties of graphene and versatile applications of graphene-based nanomaterials. The sp^2 hybridized carbon atoms, arranged in a honeycomb lattice, impart excellent electrical, thermal, optical and mechanical properties to the 2D nanostructure, named graphene. While Balandin et al predicted the high thermal conductivity (~5000 $Wm^{-1}K^{-1}$),(Balandin, 2008) Lee and coworkers have shown extremely high mechanical strength of monolayer graphene (Young's modulus of ~ 1 TPa). (Lee, 2008) In their work published in Applied Physics Letters, Moser et al demonstrated extremely high current density of monolayer graphene, which is in the order of 10^8 A/cm^2.(Moser, 2007) Moreover, monolayer graphene exhibits 97.7% transmittance. (Nair, 2008) By virtue of such properties, graphene has found its usefulness in a wide variety of application areas. However, graphene is reaching the consumer market in a variety of finished end products suitable for different applications. One of the major areas of graphene application is in the field of nanocomposites. In this chapter, we will be focused on application of graphene in polymer nanocomposites.

Nanocomposites are heterogenous materials reinforced with nanofiller additives. They exhibit enhanced mechanical, electrical, thermal and chemical properties. With rapid growth of nanotechnology, nanocomposites are also finding applications in different industrial sectors including electronics, packaging, energy, biomedicine and many more. In recent years, research in this field has been increased significantly. A similar Web of Science article search resulted in data that demonstrates rapid increment in publication within last 10 years (Figure 2a). Polymer nanocomposites are distinguished from regular composites by the fact that they have nanofiller additives incorporated in the polymer matrix. Graphene has the unique ability to be incorporated and used in versatile application fields utilizing nanocomposites, which cannot be achieved by any other nanofiller additives. Such nanocomposites are emerging as extremely useful materials featured in a stupendously growing number of publications every year (Figure 2b). According to the Web of Science, last year 3795 articles were published in different journals including the topic of graphene-based polymer nanocomposites. In this chapter, we are going to discuss application of graphene-based polymer nanocomposites, including electronics, thermal management and structural applications.

Figure 1. Rapid increment of graphene-related publications
(Source: Web of Science)

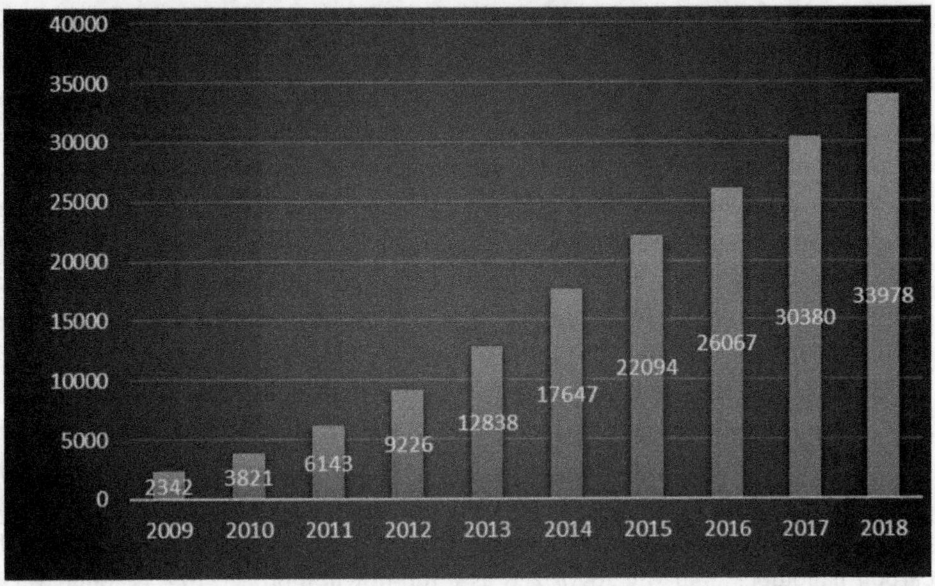

According to the Market Research Report, graphene itself does not enter the consumer market as the end product, except the research laboratories and the R&D sectors of industrial manufacturers.(Clark 2016) Till date, academic research laboratories and the R&Ds are the biggest consumer of graphene. Typically, graphene takes another form of a finished final product while being sold in the consumer market. However, manufacturing graphene has still been an extremely attractive field of study. Lately, tremendous efforts have been put into to improve the methodologies that involve large scale production of graphene, in both academic and industrial sector. In order to detail manufacturing of graphene-based polymer nanocomposites, a brief discussion on graphene manufacturing itself is warranted. There are a various review articles that are being published in peer-reviewed journals documenting the academic publications over the years. We are going to focus on the major breakthroughs, which are achieved within the last ten years related to synthetic progress of graphene production. The synthetic methodologies of nanocomposite fabrication will be discussed next, followed by discussion on different applications.

GRAPHENE SYNTHESES

Since Novoselov *et al* first demonstrated the formation of 2D graphene nanosheets (Novoselov 2005), there have been significant efforts in developing novel manufacturing methods to produce different form of graphene-based nanomaterials in larger scale. Graphene nanostructures are available in different shapes and forms, mostly depending on their method of synthesis. Apart from the monolayer, bilayer and few-layer graphene (FLG), graphene quantum dots (GQD) are also gaining attraction in the scientific community. However, 2D graphene nanomaterials are more widely used to prepare nanocomposite materials. In this section, the two most common methodologies of manufacturing 2D graphene nanosheets are going to be described.

There are two approaches, with the globally accepted terminology of top-down and bottom-up methods, currently employed to make nanostructured materials. The most commonly used top down approach involves breaking down macromolecular bulk materials to their nanodimensional counterparts. Mechanical, electrochemical and chemical exfoliation are the highly used top-down methods. Unzipping of carbon nanotubes is also categorized as a top-down approach that is found very useful in producing 2D graphene from its one-dimensional nanoform. On the other hand, the bottom-up process involves building nanomaterials from the atomic level and most commonly used methodologies include chemical vapor deposition, high pressure chemical pyrolysis, arc-discharge etc. Edwards and Coleman related these two methods with the application of graphene in their publication in Nanoscale. (Edwards, 2013) Figure 3 compares the core concept of these two methods below.

Figure 2. Increment in yearly publications related to (a) nanocomposites & (b) graphene-based nanocomposites
(Source: Web of Science)

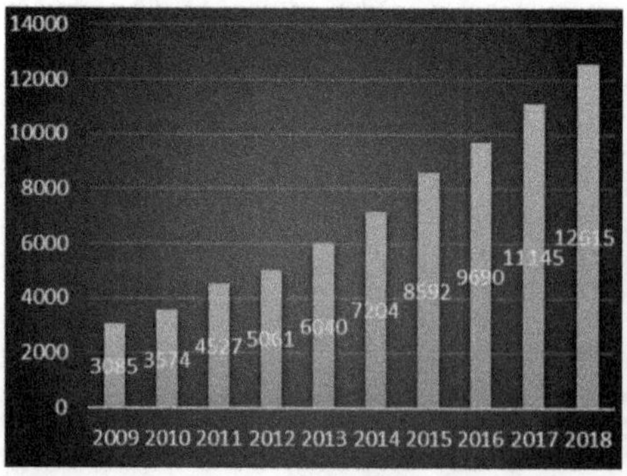

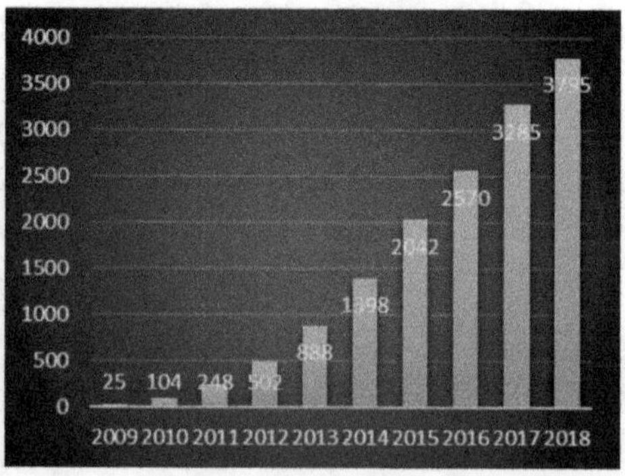

Top-down Processes

The top-down methodologies mostly rely on exfoliation techniques to prepare graphene nanostructures. Exfoliation of monolayer or few-layer graphene from their naturally occurring and readily available precursor, graphite, can take place utilizing different techniques. Novoselov and his coworkers reported micromechanical exfoliation of graphene nanosheet from graphite using a scotch tape. (Novoselov, 2005) Although this process cannot be utilized for large scale production, the method opened many pathways to fabricate graphene by peeling up layers from graphite. Apart from academic interest to investigate different properties of graphene, the micromechanical exfoliation of graphite didn't pave its way to mass scale industrial manufacturing of graphene nanostructures. However, several modifications of this process took place and have been reviewed in a variety of publications.(Rao, 2009; Huang, 2012; Coleman, 2013; Ciesielski, 2014)

One of the most popular routes has proven to be sonication assisted liquid phase exfoliation of graphene. The most commonly used organic solvents are N-methyl-2-pyrrolidone (NMP), ortho-dichlorobenzene (DCB), dimethyl sulfoxide (DMSO) and dimethyl formamide (DMF). They are chosen because of their high boiling points and higher surface tensions which keeps the graphene sheet stabilized and away from agglomeration.(Coleman, 2013) However, these solvents suffer issues related to toxicity. Bourlinos et al reported five fluorinated aromatic solvents, namely hexafluorobenzene, octafluorotoluene, pentafluoronitrobenzene, pentafluorobenzonitrile, pentafluoropyridine, to successfully exfoliate graphene flakes from graphite powder.(Bourlinos 2009) Qian et al demonstrated formation of monolayer and bilayer graphene from expanded graphite in a moderately low boiling solvent, acetonitrile.(Qian 2009) While the sonication plays a major role in graphene production via exfoliation method, the process is further assisted by addition of surfactant agents. Surfactants are organic molecules, which facilitate the dispersion and stabilization of nanosheets in the solvent medium. Surfactants are extremely

Figure 3. Schematic representation of the top-down and the bottom-up approach of graphene synthesis.
(Reprinted with permission from Ref. Edwards, 2013. © The Royal Society of Chemistry)

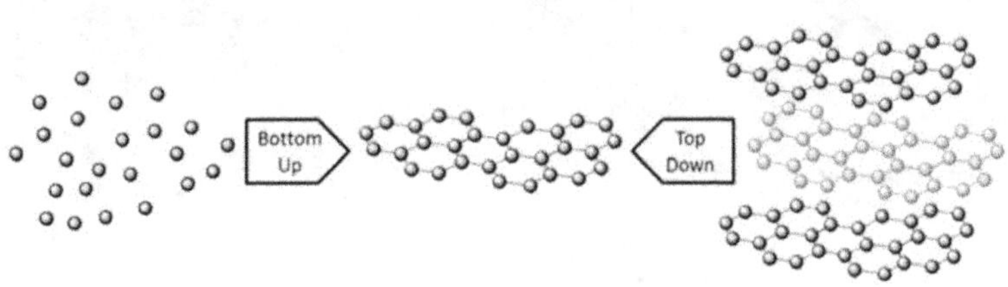

useful dispersing agents in aqueous medium. Among several surfactant molecules, polycyclic aromatic hydrocarbons are proven to be highly effective because of the π- π interaction of the aromatic ring and graphene layers.(Björk, 2010) Ciesielski et al listed the different surfactants in their review article and Figure 4 depicts the mechanism of sonication assisted exfoliation process. (Ciesielski, 2014)

Ionic liquids (ILs) are also becoming a new contender in assisting the sonication guided exfoliation process. ILs are salts in the liquid state below 100°C carrying inorganic or polyatomic organic ions. They are a green solvent and currently being used in a variety of chemical synthesis. Liu and coworkers synthesized IL-functionalized graphene nanosheets using direct electrochemical exfoliation from graphite electrodes.(Liu, 2008) Recently, Ravula et al published their work on IL-assisted exfoliation of graphene documenting investigative efforts in this area.(Ravula, 2015) Very recently, Bordes et al reported successful exfoliation and stabilization of graphene nanosheets in different ionic liquids (Table 1). Figure 5 exhibits the electron microscope images of the multilayer graphene nanosheets.

Figure 4. Liquid exfoliation of graphene with and without surfactant.
(Reprinted with permission from Ref. Ciesielski, 2014, © The Royal Society of Chemistry)

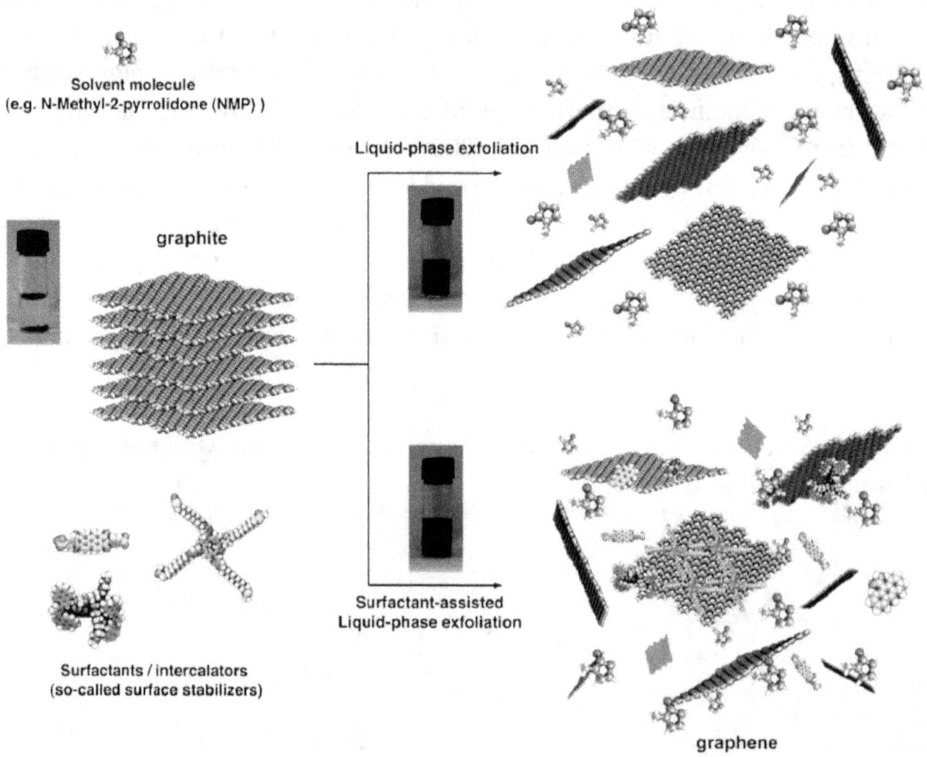

Graphene oxide (GO) has become a utility tool to produce graphene nanoflakes. The hydrophobic graphite can be oxidized to make its hydrophilic counterpart, which in turn can be exfoliated in aqueous medium. Among the several methods reported to oxidize graphitic layers, Hummers' method is the most widely used one.(Hummers, 1958) There are many modifications of the current state-of-the-art technique available for this simple process. Marcano et al reported an improved synthesis of GO by substantially changing the oxidizing agents.(Marcano, 2010) The improved method has been reported to be effective in producing large scale GO material, compared to the traditional Hummers' method. While the improved Hummers' method utilized a different ratio of $KMnO_4$, H_2SO_4 and H_3PO_4, a more economic version of this method has been published by Yu et al.(Yu, 2016) They have shown that addition of K_2FeO_4 to partly replace use of $KMNO_4$ and further reduction in the amount of H_2SO_4 can efficiently reduce the cost of GO production.

Jana et. al reported green reduction of GO using drained water from soaked mung beans (Phaseolus aureus L.). It was identified through controlled experiments and using UV-vis spectroscopy that phytic acid content in soaked mung beans is the primary contributing elements for the reduction. (Jana, 2014).

Since Stankovich et al first published the reduction of graphitic oxide to produce stable aqueous dispersion of graphene nanosheets, (Stankovich, 2006) there have been series of continuous efforts reported and reviewed.(Pei, 2012; Pumera 2013) Typically, the reduction of GO to produce chemically reduced graphene has been carried out by treatment of either hydrazine, dimethyl hydrazine, sodium borohydride followed by hydrazine, hydroquinone or UV-irradiated TiO_2. (Kim, 2010 and references within)

Figure 5. TEM images of multilayer graphene after centrifugation in (a) (C_4C_1im)(Ntf$_2$) and (b) (C_2C_1im)(Otf).

Reprinted with permission from Ref. Bordes, 2019 © 2019 Bordes, Morcos, Bourgogne, Andanson, Bussière, Santini, Benayad, Costa Gomes and Pádua

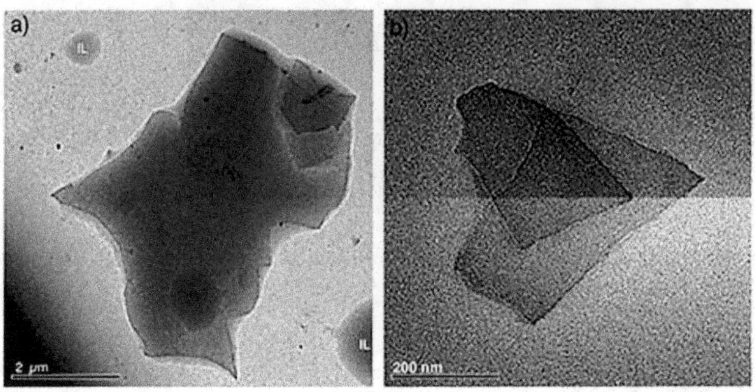

Table 1. Structure and name the ILs used for the liquid phase exfoliation of graphite.

Structure	Abbreviation
	$[C_4C_1im][Ntf_2]$
	$[C_{10}C_1im][Ntf_2]$
	$[BnzmC_1im][Ntf_2]$
	$[Pyrr_{4,1}][Ntf_2]$
	$[N_{4,1,1,1}][Ntf_2]$
	$[C_4C_1im][C(CN)_3]$
	$[C_2C_1im][N(CN)_2]$
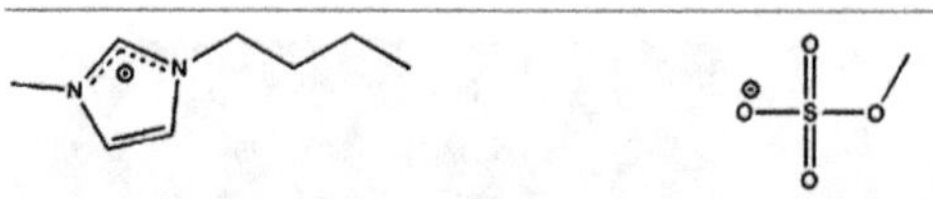	$[C_4C_1im][C_1SO_4]$
	$[C_2C_1im][Otf]$

Azizighannad and Mitra recently reported stepwise reduction of graphene oxide to form reduced GO (r-GO) and their chemical and colloidal properties.(Azizighannad, 2018) The reduction has been carried out using in situ production of hydrogen gas and they have shown in their experiments that with a reduction in oxygen content, a pronounced increment in the hydrophobicity of r-GO was observed.

Several electrochemical approaches to reduce graphene oxide producing few-layer graphene has been reported recently. The method has gained immense popularity as it can produce comparatively good quality graphene in greater yields (Toh, 2014). The graphene produced in this method is commonly called ERGO (electrochemically reduced graphene oxide). As mentioned earlier, GO is mostly produced via modified Hummers' technique. However, the electrochemical reduction of GO is carried out using different materials and solvent systems. Guo et al synthesized graphene nanosheets from graphite electrodes using -1.5 V vs glassy carbon electrode and electrochemical reduction technique shown in Figure 6.(Guo, 2009) The process is termed as the green approach since it does not require the harsh chemicals typically used for reduction process. Overall, electrochemical exfoliation is advantageous due to its simplistic approach and comparatively shorter production time.

2.2. Bottom-up Processes

As discussed earlier, the bottom-up approach is essentially a building up process, where atomic elemental carbon can grow in the form of monolayer, bilayer or few layer nanosheets. Among the different bottom-up techniques currently being employed in graphene production, the chemical vapor deposition process stands out

Figure 6. The experimental setup along with the optical images of the graphite electrode and the GO suspension before (a,c) and after (b,d) electrochemical reduction. (Reprinted with permission from Ref. Guo, 2009. ©2009 American Chemical Society)

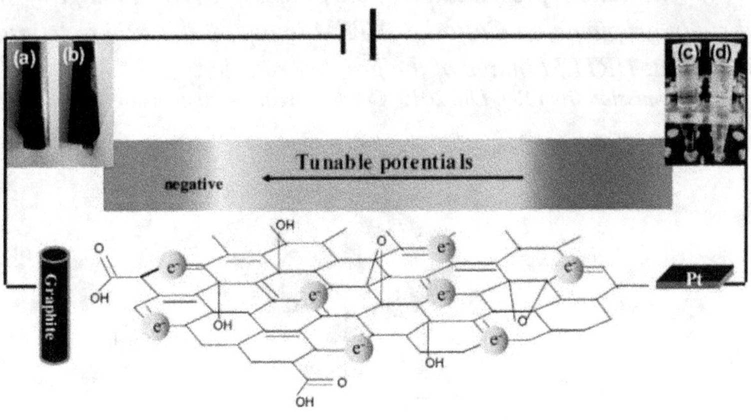

because it can make high quality graphene nanosheets. Although earlier studies have shown to use high temperature (>1300°C) annealing of SiC to produce graphene nanosheets,(Lin, 2010) typically metallic substrates are the best platforms to grow graphene. Ni, (Yu, 2008; Kim, 2009) and Cu, (Li, 2010; Li, 2011; Zhang, 2012) have been the mostly used metals for such purposes. However, growth of graphene on other transition metals including Ag, (Ayhan, 2013) Co, (Amato, 2018) and Pt (Sutter, 2009) has also been reported. In the CVD technique, the purity of the product is dependent on the surface of the substrate. Lin et el recently demonstrated production of super clean meter-scale graphene film using a Cu foil-foam stacking technique.(Lin, 2019) An electropolished Cu foil was placed under Cu foam, which in turn then kept inside a tube furnace and subsequent flow of H_2 and methane gas at elevated temperature resulted in the "super clean" graphene. Although, CVD methods require higher temperatures, with the rapid growing demand of pure and defect-free graphene nanosheets, CVD process is also finding its use in large scale graphene production.

Chakrabarti et al demonstrated a unique bottom-up approach of synthesizing crystalline graphene nanosheets.(Chakrabarti, 2011) The process involved burning magnesium metal inside dry ice and the reduction of CO_2 resulted in few layers graphene (3-7 layers, Figure 8). This technique, now commonly known as the dry-ice method, does not depend on renewable feedstocks, like graphite. Since no strong oxidizers are used in the method, it qualifies as a green technique, which has extremely low energy requirement. The method has been further modified to a self-sustaining magnesio-thermic combustion (SSMTC) process, which can be scalable to meet the large-scale production needs.(Majumdar, 2017)

Figure 7. (a) Schematic representation of the process, (b) AFM image of the freshly prepared clean graphene on Cu foil, (c) TEM image of the super-clean graphene membrane. Inset: HRTEM image of the graphene lattice.
(Reprinted with permission from Ref. Lin, 2019. © http://creativecommons.org/licenses/by/4.0/.)

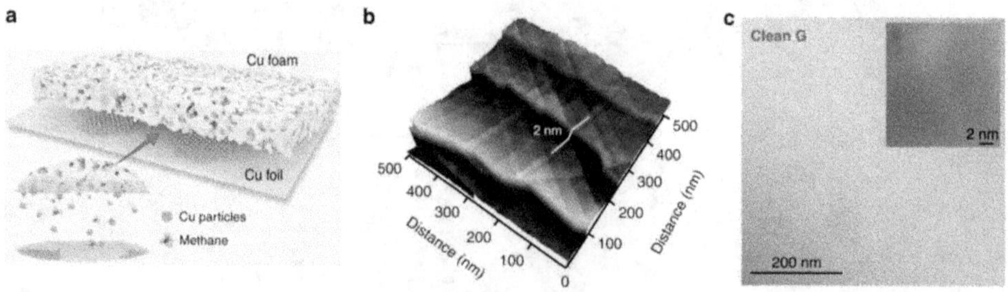

GRAPHENE-BASED POLYMER NANOCOMPOSITES

The properties of graphene-based polymer nanocomposites rely mostly on the homogenous dispersion of the nanomaterials inside the polymer matrix. The high aspect ratio of 2D nanomaterials confirms better dispersion of them as filler

Figure 8. TEM images of few-layer graphene. (a) Graphenes with an average length of 50–100 nm. (b) Larger size graphene sheets, average length 300 nm. (c) Crystalline graphenes with an average length of 200 nm. (d) High-resolution TEM image of few-layer graphenes, the number of layers ranging from 3–7. Inset: the electron diffraction pattern of graphenes.
(Reprinted with permission from Ref. Chakrabarti, 2011. © Chakrabarti et al.)

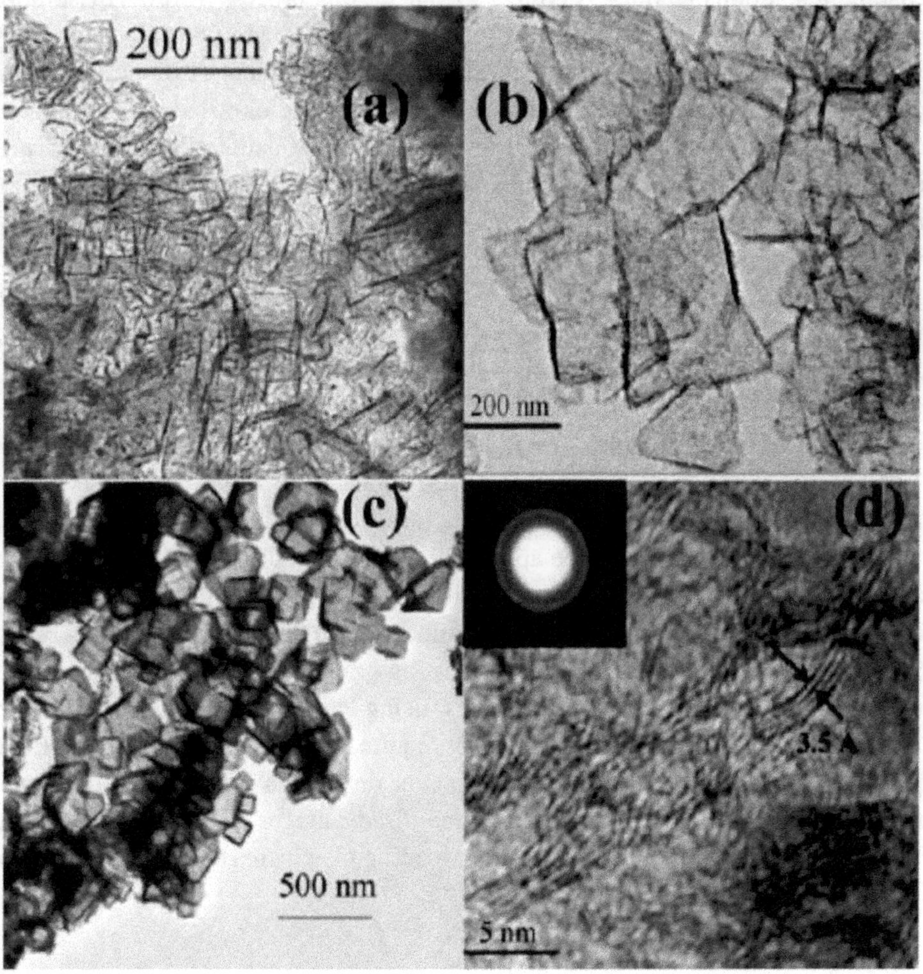

materials in the polymer matrix, which in turn ensures improvement of properties, including mechanical strength, dielectric properties and thermal conductivity. Graphene-based polymer nanocomposites have become one of the most valuable end products, which find their use in diverse application areas. While mixing graphene with other components, the excellent properties of graphene has somewhat been compromised. However, the enhancement of properties of the other component has been so pronounced that it drives the manufacturing needs of such products. In this section we are going to discuss the different manufacturing techniques of graphene-based polymer nanocomposites in general.

Solution Blending Process

The most commonly used technique to prepare a nanocomposite film is solvent casting or solution blending. A typical solvent casting process involves dissolution of the base polymer in the appropriate solvent followed by addition of the nanofiller. A well-dispersed solution is required to ensure improvement of properties of the composite films. In order to produce homogenized solution, ultrasonic dismembrators are used in different capacities. Ultrasonication time and frequency may vary depending on the nature of the polymer and particle size of the fillers. The finer the additives are, higher time and frequency are required. After a successful solution is prepared, slow evaporation of the solvent ensures formation of nanocomposite films of desired thickness. In many cases, functionalization of graphene nanosheets facilitates a homogenized dispersion of the nanomaterials in the solvent medium. Sonication plays a major role in this process as well.

Melt Blending Process

The other more environmentally benign procedure is melt-processing or melt blending, which is a solvent free approach. In this method, typically, the polymer and additives are homogeneously mixed in the molten state of the polymer and the composite is extruded or compression molded, depending on the end use. Figure 9 shows a representative scheme of fabricating graphene-based linear low density polyethylene (LLDPE) nanocomposites using extrusion technique.(Khanam, 2016) The melt blending process does not require any solvent usage; however, any pretreatment of graphene may ensure a better dispersion of the nanofiller inside the polymer matrix. The process has some disadvantages as the amount of filler loading is typically restricted. The higher loading may increase the viscosity of the melt and hinder the processing part.

In Situ Polymerization Process

The other way to prepare the polymer nanocomposites is via in situ polymerization. In this process graphene nanofiller is loaded and dispersed in the liquid monomer and the polymerization takes place at a desired temperature and time. This process may use a solvent sometimes, which is a major disadvantage as the solvent removal is difficult after polymerization. Dispersion of graphene in the monomer can become challenging and the amount of filler loading is also restricted. Higher filler loading may increase viscosity of the liquid phase, which is not desired for these applications.

So far, we have discussed the general procedures for manufacturing graphene-based polymer nanocomposites. Each method has their own advantages and disadvantages. They require different conditions and processing parameters including processing time and temperature. In the following segment we will briefly discuss the versatile utilization of graphene-based nanocomposites in mechanical, electrical and thermal applications.

APPLICATION OF GRAPHENE-BASED POLYMER NANOCOMPOSITES

Graphene-based nanocomposite materials find their use in diverse applications since graphene itself can act as multifunctional additive. The excellent properties of graphene enhance the properties of the polymer composites in many ways. The

Figure 9. Schematic representation of nanocomposite fabrication via extrusion.
Reprinted with permission from Ref. Khanam, 2016. © 2016 Elsevier Ltd.

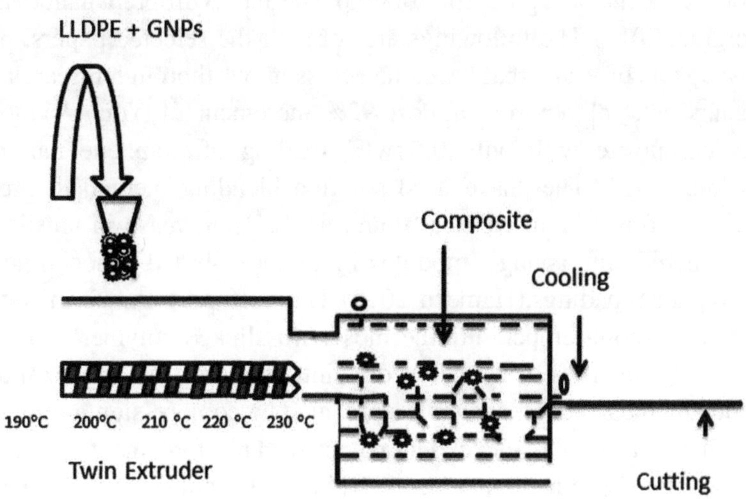

mechanical strength of the composite films has shown dramatic improvement in numerous occasions. Electrical and thermal conductivity of the composites have also demonstrated similar trends. We will review and cite various references to portray such properties of graphene-based nanocomposites. A variety of polymers are used depending on the use of the finished product. Many a times, the process employed to make such composites solely depends on the choice of the polymer. The major highlight of this section will be polymer used along with the manufacturing techniques. We have categorized this section in three categories depending on the final application.

There are a few comprehensive reviews available on properties and application of graphene-based nanocomposites.(Kim, 2010; Kuilla, 2010; Dhand, 2013) However, as we have discussed earlier, number of publications in this field keep on rocketing at a great pace. In our review, we would like to focus on the very recent trends of publications. Many literatures have shown multiple property enhancement in their respective publications. We may have used the same references in different segments as we found them appropriate.

Nanocomposites With Enhanced Mechanical Properties

Incorporation of graphene in polymer matrix certainly improved the mechanical strengths of the composites. In early 2005, Mack et al reported that graphitic nanoplatelet reinforced polyacrylonitrile-based nanocomposites showed two-fold increment in its normalized Young's modulus. Till date, there are many such findings available in published literatures. Improved mechanical strengths are pronounced in enhanced mechanical properties, including Young's modulus and tensile strength. Papageorgiou et al published an excellent thorough review in the Progress in Materials Science on the mechanical properties of graphene and its different nanocomposites. (Papageorgiou, 2017) The followings are some of the selected papers, published within last couple of years, that we came across in our thorough research.

Berhanuddin et al reported almost 92% increment of Young's modulus in an epoxy composite with only 0.5 wt% loading of graphene nanoplatelets. (Berhanuddin, 2017) They have used solution blending method to prepare the nanocomposite films. Hameed et al found an 125% increase in tensile strength and a 21% increase in Young's modulus in an epoxy-based nanocomposite with 0.6 wt% graphene loadings.(Hameed 2018) Their effort is unique in way as they have prepared a room-temperature thermoset crosslinked polymer with very high flexibility. The properties were achieved by introduction of an imidazolium based ionic liquid and the nanocomposite was prepared by compression molding. Young et al have shown in detail that excellent mechanical reinforcement can be achieved by dispersing graphene nanoplatelets (GNPs) in thermoplastic elastomers (TPE) and polypropylene (PP) (Young, 2018). The nanocomposites were fabricated via melt processing using a twin-screw extruder at temperatures varying from 165°C (for TPE) to 190°C (for PP). Aoyama and coworkers have modified graphene using trimellitic anhydride group.(Aoyama 2018) The modified graphene was successfully employed to manufacture a polyethylene terephthalate-based nanocomposite via melt processing. The modified graphene composite graphene exhibited about 5% higher Young's modulus than its PET/graphene composite counterpart (Figure 10). A very recent article published by Kumar et al produced excellent mechanical properties of a variety of nanohydroxyapatite (nHAp) and graphene nanoparticles (GNPs)-based composites. (Kumar 2019) They have reported significantly higher compressive strength and elastic modulus of the nHAp–GNPs composites than the pure polymer and such has been attributed to the role of the bonding interface between nHAp and GNPs.

Nanocomposites With Enhanced Electrical Properties

The extremely high electrical conductivity and very high carrier mobility made graphene a unanimous choice for many electronics applications. Consequently, graphene-based composites are finding their usefulness in similar fields of applications. In this section we are going to review some of the advancements of nanocomposites in this direction.

Application of graphene nanostructures as ultracapacitor was first identified by Stoller et al. (Stoller, 2008). Si and coworkers have described a process of making

Figure 10. Functionalized graphene and enhanced mechanical properties of the composite with PET.

(Reprinted with permission from Ref. Aoyama, 2018. ©2018 American Chemical Society)

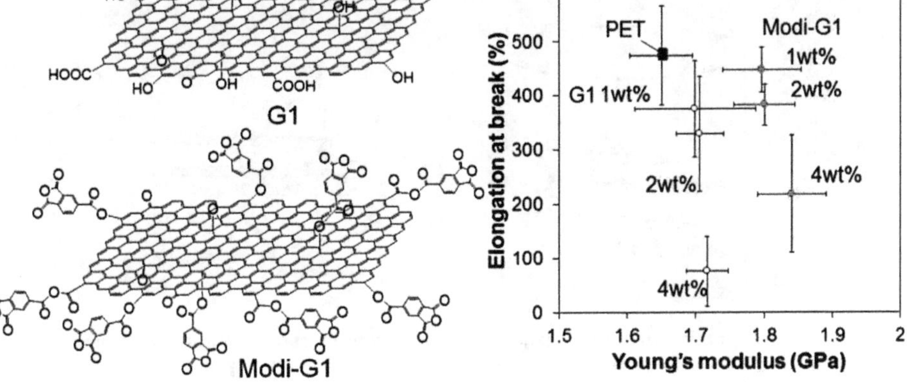

graphene-platinum nanocomposites using liquid exfoliation technique (Fig. 11) and the Pt-graphene composite exhibited a very large capacitance of 269 F/g. (Si, 2008) Yu et al reported moderately high capacitance of 120 F/g for a graphene, carbon nanotube hybrid composite using poly(ethyleneimine) polymer. (Yu, 2009) The polymer composite was prepared via solution casting method over the indium tin oxide (ITO) glass substrate.

Mishra et al reported graphene-based supercapacitor nanocomposite with polyaniline (PANI).(Mishra 2011) Few-layer graphene was synthesized using hydrogen exfoliation techniques, which involved thermal exfoliation of graphitic oxide under hydrogen atmosphere. PANI-f-HEG nanocomposite was fabricated utilizing polycondensation of the monomer. They have also successfully prepared functionalized graphene coated metal oxides (Fig. 12) and compared the capacitance of the different electrodes at varying scanning rates. Fe_3O_4, RuO_2 and TiO_2 based nanocomposites were synthesized and characterized along with graphene-based PANI nanocomposites (PANI-*f*-FEG). PANI-*f*-FEG has shown the highest capacitance of

Figure 11. Schematic of graphene sheets and nanoparticle-modified graphene sheets in its dispersion and dry state.
(Reprinted with permission from Ref. Si, 2008. © 2008 American Chemical Society)

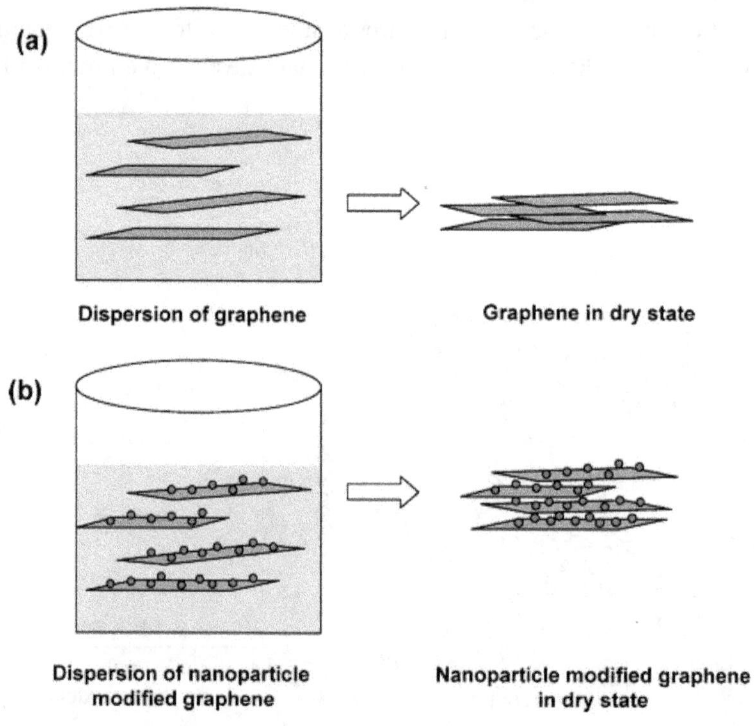

375 F/g at the 10 mV/s scan rate, which can be attributed to the conductive nature of PANI itself.

A graphene – polypyrrole (G/PPy) composites was prepared via in situ polymerization technique and its electrochemical impedance was investigated for potential application as electrochemical supercapacitor electrode.(Basnayaka, 2013) The nanocomposites exhibited a very high specific capacitance of 256 F/g at the discharge current density of 0.5 mA/g . The porous structure of the composite materials was found to be beneficial for the specific application of the nanocomposite as electrodes. More recently, Yang et al produced a highly conductive graphene-polyvinyl alcohol composites using the traditional solvent casting method. (Yang, 2018) The graphene sheets were slightly oxidized to facilitate dispersion of the filler in the polymer matrix. The nanocomposite showed the conductivity of 25 S m^{-1} with a 6.25% loading of the oxidized graphene sheets. The very high conductivity values translated to be superior over any available graphene- and GO-based nanocomposites.

Figure 12. TEM images of HEG (a), RuO_2-f-HEG (b), TiO_2-f-HEG (c), Fe_3O_4-f-HEG (d), and PANI-f-HEG (e) nanocomposites, respectively.
(Reprinted with permission from Ref. Mishra 2011. © 2011 American Chemical Society)

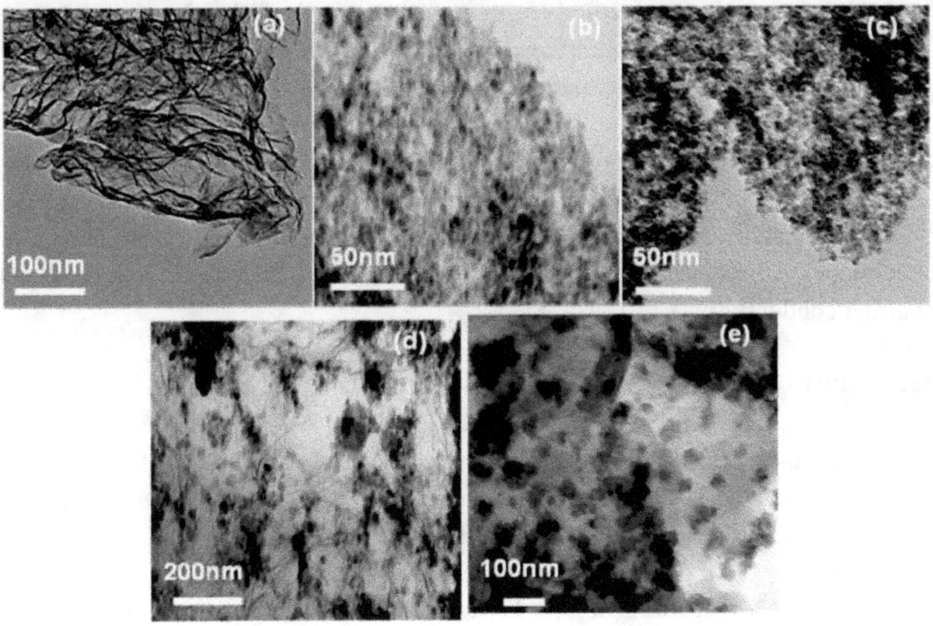

Nanocomposites With Enhanced Thermal Properties

The superior thermal conductivity of graphene warranted its application in several thermal management systems and as thermal interface materials. Packaging and electronic industries are also among the beneficiaries of thermally conductive graphene-based composites. Since Balandin published their work on the thermal conductivity, (Balandin, 2008) there have been numerous efforts on utilizing this property of graphene towards the applications we just mentioned. In this section we will document some of these recent endeavors.

Yavari et al investigated stearyl alcohol and graphene-based nanocomposites as a nanostructured phase changed material. (Yavari, 2011) With a very minimal loading of (4 wt%) graphene they have prepared the composite utilizing the solvent casting method. The product has shown 2.5 times increment in thermal conductivity than the base polymer. Khanam et al prepared composites with linear low-density polyethylene (LLDPE) and graphene nanoplatelets.(Khanam, 2016) We have shown their processing scheme earlier in Fig. 9. In their experiments, they have established a correlation between the extruder speed and properties of the nanocomposites. The highest thermal conductivity of ~ 0.5 W/mK was achieved by the composite prepared with 10 wt% graphene using a 150-rpm extruder speed. The higher speed contributed towards better dispersion of the nanofiller and prevented them to agglomerate in the polymer matrix.

Majumdar and Chakrabarti (Majumdar, 2017) reported polydimethyl siloxane (PDMS) and epoxy-based nanocomposites with the SSMTC fewlayer graphene using the solvent casting method. As demonstrated in the Table 2, we can observe more than 10-fold increment of thermal conductivity of the nanocomposites when compared to their base polymeric counterparts. It is also worthy to note that, in the case of PDMS composites the percent loading of the nanofiller is only 2 wt%.

Wang et al fabricated polyamide 6 (PA6) nanocomposite with graphene using in situ polymerization. (Wang, 2018) The 3D network composite structure has shown thermal conductivity of 0.69 W/(m. K), with only 0.25 wt % graphene. The increment is registered to be 188% compared to that of pure PA6, which shows thermal conductivity 0.24 W/(m. K).

Graphene Composite for Thermal Interface Material

Power demands of many of the electronics devices has been consistently increasing along with the need for an efficient thermal management system to meet the demand for dissipating the high heat flux losses and reduce the junction temperature to sustain device operation. Air gaps that exist between the heat source and heat sink/cold plate increase the resistance to heat flow between them. Excessive heat accumulated in

Table 2. Thermal conductivity data for PDMS(S6)-SSMTC-FLG and LO Epoxy(UV-30) -SSMTC-FLG composites.

Sample	Pressure (psi)	Temperature (°C)	Thermal Conductivity (W/m. k)
PDMS S6 +2% SSMTC-FLG (Al Face sheet, t=6.56mm)	40	24.2	2.761
	40	52.55	2.838
PDMS S6 +2% SSMTC-FLG (Al Face sheet, t=6.41mm)	40	28.75	2.132
	40	56.72	2.224
PDMS S6 (Al Face sheet, t=6.40mm)	40	28.25	0.246
LO Epoxy UV-30 (Al Face sheet)	40	20.95	0.202
	40	50.48	0.255
LO Epoxy UV-30 + 10% SSMTC-FLG (Al Face sheet)	40	22.75	2.356
	40	51.7	2.454

electronics system reduces performance and causes wear and premature failure of the system. High power density electronics components have specifically called for the reduction of the thermal resistance between power module components and heat sink/cold plates by a sizable factor. Thermal interface material (TIM) is used to fill the air gaps to reduce the resistance between the heat powder density device and heat spreader or carrier like heatsinks or cold plates. Development next generation of Thermal Interface Materials of high thermal conductance is in demand for use in high powder density applications.

Design and development work are performed at Northern Illinois University (NIU) (Syamala (2016), Varadaraju (2017)) in an effort to design advanced TIM using graphene-metal, graphene-polymer, graphene-phase change material or nanotubes-metal composite that can outperform traditional TIMs when it is used in high heat fluxes applications. The main aim of this research is to develop advanced thermal interface material using different structure of filler matrix composition, mixture composition and thickness of number of candidate samples. Generally, TIM layer has the maximum thermal resistance in the package. Reducing the thermal resistance of this layer leads to enhanced heat dissipation from heat source to the heat sink, and thereby decreasing maximum temperature in the package.

TIMs should have high thermal conductance, good capability to fill up the micro gaps, and good thermal stability over wide range of thermal conditions. TIM placed in between the heat source and heat sink significantly increases the heat transfer capability of the system. However, with the use high thermal conductivity composite system of TIMs, additional resistance to heat transfer exists at the contact interface of dissimilar materials, and this causes temperature drop across the interface as depicted in the figure below.

Thermal resistance of the TIM depends not only on the thermal conductivity, but also on the thickness of the BLT. The BLT associated with TIM varies with type and volume fraction of the particles and the applied pressure. The effective thermal conductivity of the TIM has two components: one associated with the bulk resistance of the TIM and second associated with the contact resistances of the TIM with the two substrate interfaces, R_{c1} and R_{c2} at the top and the bottom respectively.

Improper selection of TIM in terms of *bond line thickness (BLT)*, melting point, reliability, material properties, etc., increases the interfacial thermal resistance and contact resistance. Conventional TIMs such as wax, grease, thermal tapes, gels and pads face challenges to satisfy the above-mentioned requirements due to decrease in size and increased speed, reliability of the new generation of electronics and power modules systems. High thermal conductivity and low thermal resistance of TIM is achievable by dispersing high-conductivity filler material in a polymeric base or matrix materials. Major factors that contribute to the reduction of the thermal resistance of the TIM are i. increased thermal conductivity of TIM; ii. reduced Bond Line Thickness (BLT); iii. reduced contact resistances (R_{c1} and R_{c2}).

Advanced Thermal Interface Materials Based on Graphene

There has been interest in maturation of nanostructure-based materials like nanotubes, graphene, diamond, boron arsenide technology and material systems. In that line, the intent is to develop more reliable thermal interface materials that should exhibit few orders of magnitude enhancement in effective thermal conductivity specially to meet the cross-plane effective thermal conductivity enhancement of the order of 500% by meeting the goals of 20 to 25 W/(m. K).

Common TIMs are the composite mixture of thermally conducting metallic or ceramic fillers and compatible polymeric materials. Major requirements are

Figure 13. Thermal contact resistances and temperature variation at the TIM-Filled interface of heat generating module and heat sink.

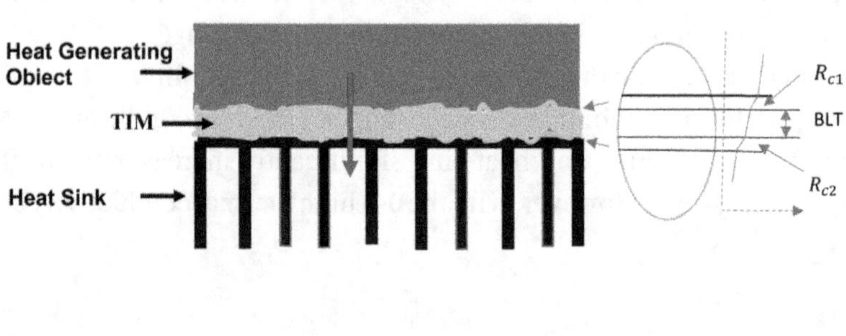

high thermal conductivity or thermal conductance and compliance. Such common TIMs require large volume fractions of fillers. Polymeric composites with a high thermal conductance are always desired for wide variety of applications. While improved thermal conductivity of polymers can be achieved through dispersion of metallic particles in a polymer matrix, a good dispersion and thermal coupling is quite a challenge. Thermal resistance and reliability of the interface material must be improved significantly for effective use in high temperature and high heat flux electrical power systems.

In recent years, major research effort on the development of high thermal conductivity thermal interface materials. Carbon nanotubes (CNTs- SWNTs), boron nitride nanotubes – high-K BNNTs, boron nanotubes (BNTs) and graphene –bilayer and few layer graphene (FLGs) are some of the novel materials that have been considered for the development of new generation TIMs with high thermal conductance or reduced thermal resistance. In previous efforts to develop thermal interface materials (TIMs) based on carbon nanotubes (CNTs) suffered from Kapitza or thermal contact resistance effects (Hansson et, al (2016) and failed to provide the anticipated performance. It is expected that the effective thermal conductivity of CNT-based TIM is significantly less that the individual CNT. However, major reason for much lower value is in the lower pack density and imperfect orientation CNTs and imperfect contacts at the mating surfaces. The graphene nanoplatelets were the early versions of graphitic graphene with 10 to 20 layers. Recent reports from Balandin *et al.* showed that graphene with appropriate aspect ratio can help in reduction of Kapitza resistance and promote thermal coupling between graphene filler and matrix materials to increase effective thermal conductivity.

In order to reduce interfacial resistance at the CNT ends, the CNT- polymer composite has been considered by Marconnet et. al (2011).They investigated the impact of aligned MWCNT density on the thermal conductivity of nanocomposite. MWCNT arrays were grown using Chemical Vapor Deposition (CVD techniques and densified mechanically to increase density. Infrared (IR) microscopy was used to characterize axial and transverse thermal conductivities of the CNT-polymer composite. A free-standing film is formed by infiltrating an MWCNT array with thermoset- epoxy polymer. Cola et. al proposed growing double-sided CNT arrays on metal foils to improve contact interfaces. (Cola, 2007)

Graphene has very high in-plane thermal conductivity and significantly higher than most of the traditional thermal interface materials. It enhances the heat dissipation evenly in all in-plane directions. However, like any other thin film, the cross-plane thermal conductivity of graphene is order of magnitude less than the in-plan thermal conductivity. The cross-plane thermal conductivity varies significantly with the thickness or number of graphene layers. Graphene is widely used as a filler material in composite materials due to its outstanding thermal properties as well as lower for cost compared to other materials. Graphene-metal, Graphene-polymer, Graphene-Phase change material or nanotubes- metal composites can outperform traditional TIMs when they are used in high heat fluxes. The thermal conductivity thermal conductance of the composite thermal interface materials purely depends on the types of composition and types of filers and base materials. Various composites such as graphene-metal, graphene-phase change material, graphene-polymer or nanotubes-metal composites are being investigated.

Ghosh et. al (2010) measured the intrinsic thermal conductivity of few layer graphene sheets by increasing number of atomic planes. They studied that at room temperature this intrinsic thermal conductivity changes from 2800 to 1300 W/(m. K) as the number of atomic planes increases from 2 to 4. Balandin (2011) reviewed the thermal properties of various carbon materials including graphene, carbon nanotubes, graphite nanoplatelets and graphene flakes. The increase in thermal conductivity is very high for the base material when carbon materials are used as fillers compared to the traditional fillers. Li et. al (2009) discussed in their paper synthesized large area graphene films by chemical vapor deposition technique using methane on copper substrates. Seol et. al (2010) presented in their paper that they fabricated monolayer graphene paper exfoliated on a silicon dioxide. Thermal conductivity of this material is 600 W/(m. K) which is higher than most of the metals (4). Wu and Drzal (2012) fabricated new flexible graphene paper using graphene nanoplatelets. The thermal conductivity of this graphene paper is around 313 W/(m. K). They observed that when this graphene paper is inserted into multilayer composites it enhanced the in plane thermal conductivity of the composite.

Shahil and Balandin (2012) in their paper described the thermal properties of graphene and few layers graphene in thermal management applications. They indicated that graphene based thermal interface materials performed better than the CNT because they have lower ***kapitza*** resistance at the interface. Park et. al (2015) developed a high performance FLG composite thermal interface material using exfoliation method. They measured the thermal interface resistance between FLG and copper layer. Narumanchi et. al (2008) in their paper measured the thermal resistance of various thermal interface materials by using ASTM D5470 test method designed based on steady state method.

Few Layer Graphene (FLG) and Graphene Composites

Graphene is two-dimensional a monolayer graphite with over 100-order anisotropy of heat flow between the in-plane and cross-plane directions. Graphene possesses one of the highest in-plane thermal conductivity of the order of 2000-4000 W/(m. K) for suspended graphene. In comparison thermal conductivity of natural diamond is 2200 W/(m. K) at room temperature (Pop et. al (2011)). The high in-plane thermal conductivity is due to covalent bonding between carbon atoms and the out-of-plane conductivity is due to the weak van-der-waals coupling force. The in-plane thermal conductivity of graphene, however, decreases considerably when supported with a substrate or encased/confined due to weak thermal coupling caused by van der Waals interactions and interaction/scattering of graphene phonon with the substrate phonons. Thermal conductivity of graphene supported by silicon substrate is reported as 600 W/(m. K).

Thermal properties like specific heat and thermal conductivity of nano- or micro-structured materials are linked to the phonon transport phenomena involving multiple acoustic and optical vibration modes of lattice structure. A graphene unit lattice cell structure is subjected to three acoustic modes and three optical phonon modes. The Theoretical modeling of thermal conductivity requires consideration of solid-state physics and consideration dispersion curves, a relationship between the phonon energy, E or frequency, ω and phonon wave or group velocity.

Thermal conductivity graphene also decreases as it stacks into layers like in few layer graphene (FLG) which are then subjected to cross-plane heat dissipation with van-der-waals interaction and phonon scattering in between layers (Balandin and Nika (2015)). Singh et al. (2011) have demonstrated their theoretical analysis based on phonon-phonon interactions and linearized Boltzmann transport equation that out-of-plane thermal conductivity dropped sharply from a single-layer graphene to two-layer graphene. With further increase of layers to three or four, the decrease is less with increase in layers and remains unchanged by four layers. At about 300K, the thermal conductivity (about 3000 W/m. k) dropped by 29% for two layers to 37% to 35% and 37% (2000 W/m. K) for three layers and four layers respectively.

Few layer graphene (FLG) and graphene composites are also considered increasing as a viable improved low thermal resistance interface material. Few layer graphene (FLG) Composite of optimized mixture graphene as the fillers and a range of matrix materials such as metallic micro – and nanoparticles are considered at NIU. Results for enhanced thermal conductivity by a factor of ~ 23 is achieved with commercially available TIM mixed with 10% loading of graphene as thermally conductive filler.

MWNT- graphene nanoplatelet composite (20% MWNT-80% graphene nanoplatelets) was also considered. Graphene nanoplatelets are the early versions of the graphite graphene with 10 to 20 layers. Major huddle in development and commercialization of graphene-based nanocomposites was the high expense of graphene production due to the restriction on the base starting material-graphite, which is limited in resources. Newer low-cost and large volume graphene production capabilities will create game-changing opportunities for the usages of graphene and graphene composites.

Experimental Characterization Techniques

Among different steady-state techniques, absolute technique is commonly used for the measurement of bulk material such as composite thermal interface materials. Transient technique is often used for thin-film nano-structured materials. Measurement of thermal properties of nano-structured materials such as NT and Graphene are quite challenging. Measurement techniques often classified based on the way the sample is heated such as i. experiments that uses external heating to create temperature gradient and ii. experiments that uses self-heating of nanotubes by applying voltage difference across the sample and create temperature gradient from resulting electrical resistance changes.

In 3ω-thermal conductivity techniques (Cahill (1990)) metal heater is used on one side of the sample to apply period heating and the oscillations are measured to estimate thermal diffusivity and thermal conductivity. Measurement techniques such as micro-heating and 3ω for measuring thermal conductivities face challenges due to error associated with the interfacial contact resistance. Transient technique such as Laser Flash Technique (LFT) is a very fast and a precise way of measuring thermal diffusivity and thermal conductivity over a very wide range of temperature band. Newer techniques such as Raman thermometry techniques (Doerk et. al (2010), Sansoz (2011), Soudi et. al (2011)) is increasingly being used as a nondestructive thermal characterization technique for graphene and NTs. This technique also involves independent measurement of optical absorption at specific wavelength based on Maxwell's equation.

Researchers at NIU worked on the development of polymer-graphene composites thermal interface applications (Syamala (2016), Varadaraju (2017)). The polymeric materials used for the nanocomposite formulation are processed with appropriate nanofillers developed and designed in our laboratories. The PDMS silicone, and the cycloaliphatic epoxy-based resins are commercially available and they were chosen as they meet the out-gassing requirements. Solution casting method was used to produce the nanocomposites. A solution was prepared using the nanomaterials and the polymer matrix in suitable solvents, which was then casted on an aluminum substrate and dried under vacuum to produce the nanocomposites. A variety of polymer-graphene composites TIM samples with varied graphene filler composition and base materials were investigated A steady-state absolute thermal conductivity measurement set-up was designed and built following ASTM D5470 test method and several polymer-graphene samples were tested.

Performance Evaluation Using Simulation Analysis

A three-dimensional computational analysis model is developed to evaluate number graphene-based TIM in terms of spreading of heat and reduction in junction temperature using power device that uses Insulated-Gate Bipolar Transistor (IGBT) power module as shown in the figure below.

Computational analysis can be performed to evaluate different material composition for TIM for high heat flux applications. Different FLG sample materials are evaluated and compared with the state-of-the art thermal interface materials such as thermal grease or gel. Some sample comparison results are presented with the following quantities: maximum temperature in the module; contour plots of temperature distribution at different planes; line plots of temperature, drawn at the bottom of IGBT in the Z-direction, and heat flux plots to quantify effectiveness of the FLG - based TIM in spreading and in lower temperatures at the contact interfaces.

Figure 15 show the temperature variation over the power generating module. Figure 15a for the use with thermal grease show that the maximum temperature in the module is 132.77 °C which is right at the IGBT. The temperature gradually decreases as heat spreads away from the IGBT and Diode heat sources towards the edges and to the heat sink. The minimum temperature in the module is 90.968 °C. Figure 15b for the case of Dow Corning TC 5022 shows improvements. The maximum and minimum temperatures in the module are now reduced to 91.951 and 70.421 °C respectively.

Figure 16 shows temperature distribution at the top and bottom different planes of the TIM layer.

Figure 14. Assembly of IGBT module

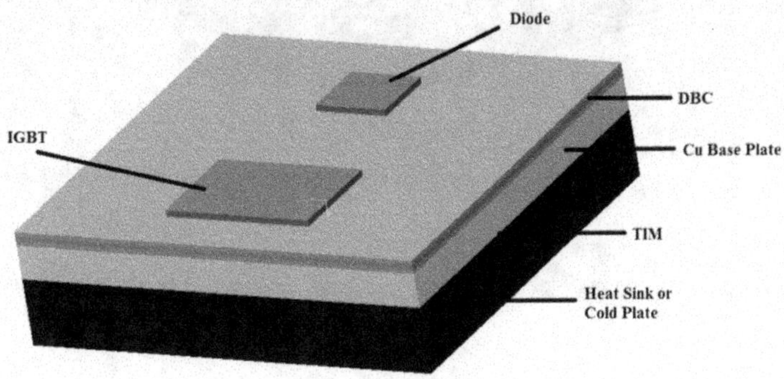

Figure 15. Temperature variation within the module

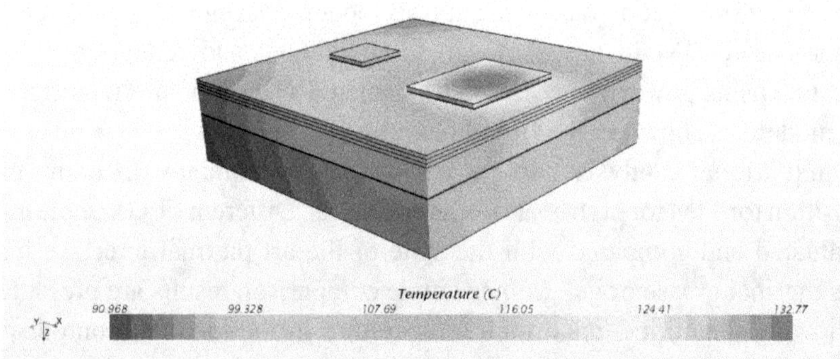

(a) Thermal grease

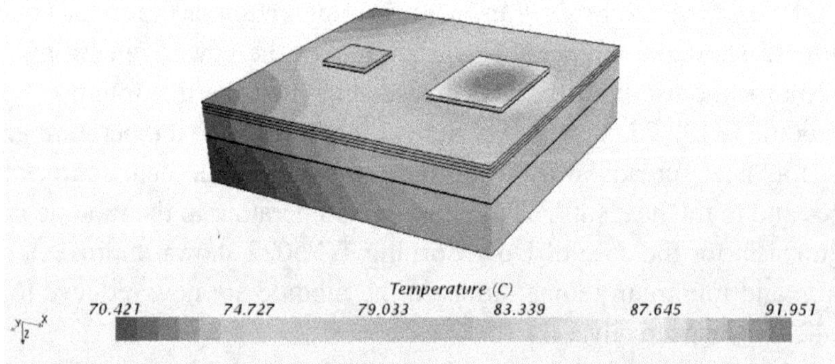

(b) Dow Corning TC 5022)

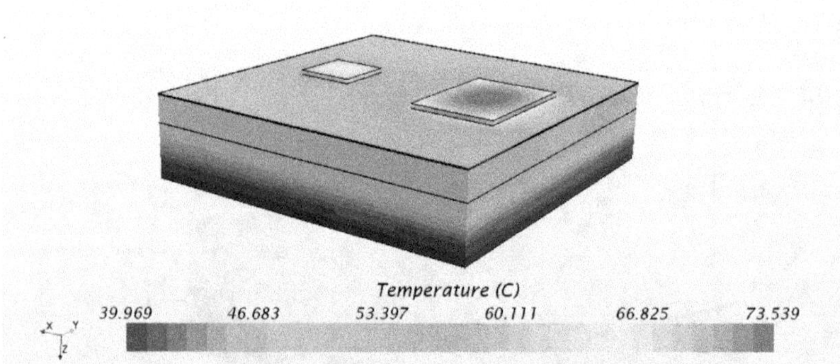

(c) FLG: Kx = Ky = 1500, Kz = 10 and thickness = 100 μm

The difference between the maximum temperatures in the two layers of the FLG layer is reduced significantly to a 0.321 °C in comparison to 3.118 °C and 21.51 °C for thermal grease and Dow Corning TC 5022 respectively.

Figure 17 shows the line plots of heat flux variation at different horizontals planes: top of DBC, and top and bottom layers of the TIM.

Results show the variation of heat flux along the three horizontal line probes drawn in the XY plane. Heat flux is more at the spots where diode and IGBT located for the top DBC layer. As we go from DBC layer to the TIM layers, the heat flux variation is significantly reduced for the case with FLG layer as compared to the Thermal Grease and Dow Corning TC 5022. FLG layer is spreading the heat more effectively in cross plane (XY) thereby decreasing the maximum temperature in the module.

Figure 16. Temperature distribution at the top and bottom planes of the TIM layer

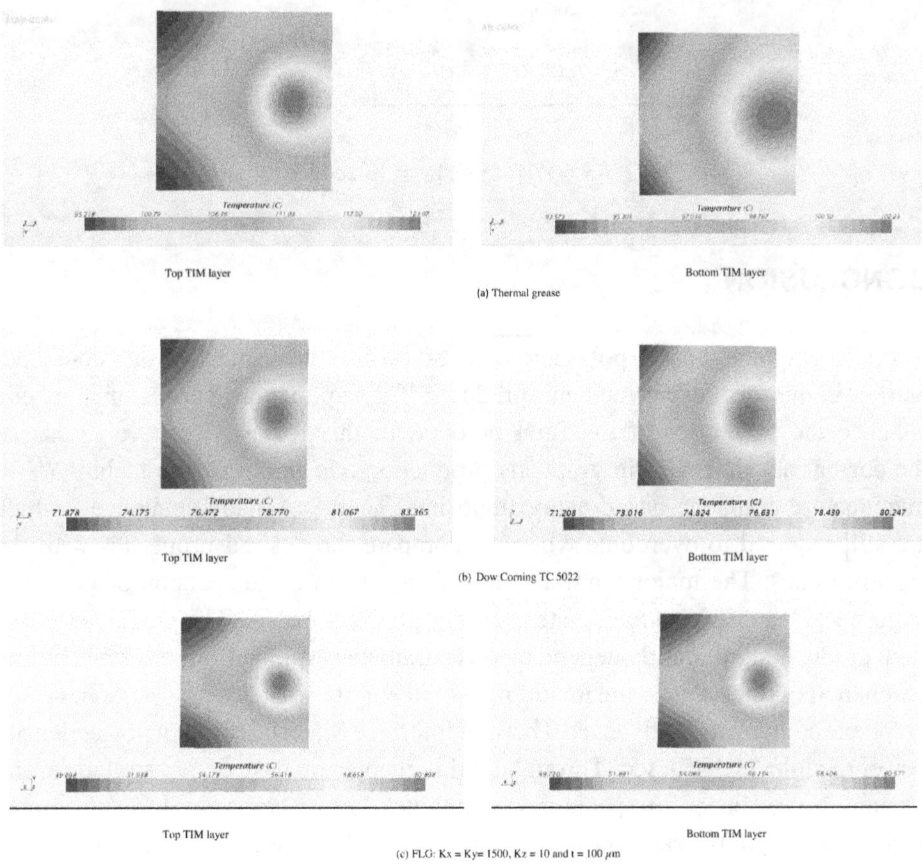

Figure 17. Heat spreading along XY- horizontal line

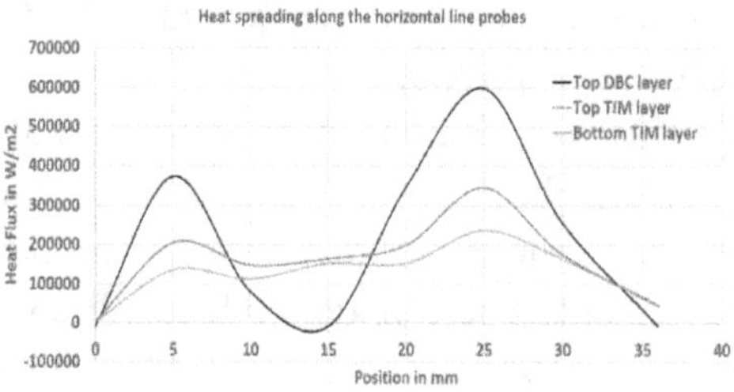

(a)Thermal grease

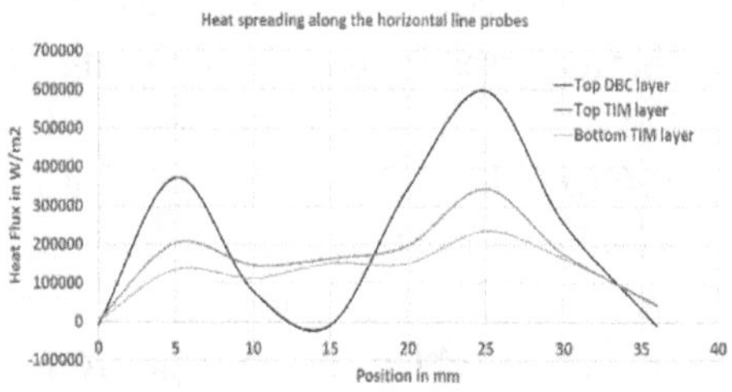

(b)Dow Corning TC 5022

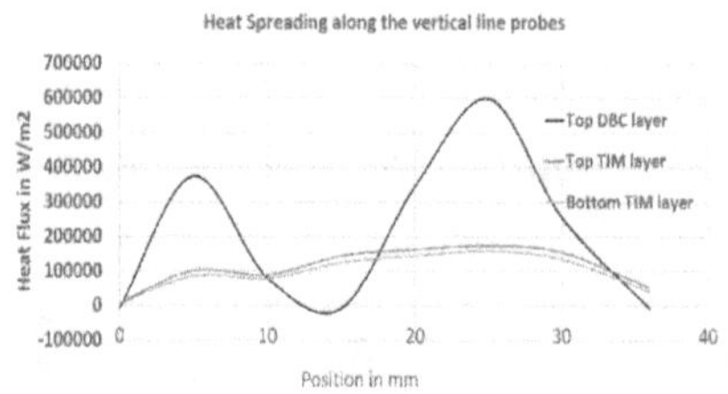

(c) FLG: Kx = Ky= 1500, Kz = 10 and t = 100 μm

CONCLUSION

Overall, graphene-based polymer nanocomposites are entering the consumer market in the form of various end products. The superior properties of graphene enhance the properties of the final product. In this review, we have discussed the current advancement in graphene production and the available techniques to manufacture graphene-based nanocomposites. There are many challenges, which are still required to overcome when we compare the desired properties with the achieved ones. The major hindrance has always been the dispersion of graphene nanosheets /nanoflakes / nanoplatelets in the polymer matrix. The properties of the final products significantly depend on this parameter. Applications of graphene and graphene-polymer composite including the potential use of few layer graphene and graphene composite as advanced Thermal Interface Material for high power density power modules is discussed. Throughout this chapter we have demonstrated several commonly used or certain specialized techniques that are being employed to combat such issues. With the rapid progress of science and technology, particularly in the field of nanomaterials, we can always be hopeful about meeting all the targeted accomplishments soon.

REFERENCES

Amato, G. (2018). High Temperature Growth of Graphene from Cobalt Volume: Effect on Structural Properties. *Materials, 11*(257), 1-14.

Aoyama, S., Ismail, I., Park, Y. T., Yoshida, Y., Macosko, C. W., & Ougizawa, T. (2018). Polyethylene Terephthalate/Trimellitic Anhydride Modified Graphene Nanocomposites. *ACS Appl. Nano Mater.*, *1*(11), 6301–6311. doi:10.1021/acsanm.8b01525

Ayhan, M. E., Kalita, G., Sharma, S., & Tanemura, M. (2013). Chemical vapor deposition of graphene on silver foil as a tarnish-resistant coating. *Rapid Research Letter*, *7*, 1076–1079.

Azizighannad, S., & Mitra, S. (2018). Stepwise Reduction of Graphene Oxide (GO) and Its Effects on Chemical and Colloidal Properties. *Scientific Reports, 8*(10083), 1-8.

Balandin, A. (2011). Thermal properties of graphene and nanostructured carbon materials. *Nature Materials, 10*(8), 569–581. doi:10.1038/nmat3064 PMID:21778997

Balandin, A. A. (2015). *Thermal Transport in Graphene and Graphene Multilayers*. Riverside, CA: University of California, Moldova State University.

Balandin, A. A., Ghosh, S., Bao, W. Z., Calizo, I., Teweldebrhan, D., Miao, F., & Lau, C. N. (2008). Superior Thermal Conductivity of Single Layer Graphene. *Nano Letters*, *8*(3), 902–907. doi:10.1021/nl0731872 PMID:18284217

Basnayaka, P. A., Ram, M. K., Stefanakos, L., & Kumar, A. (2013). Graphene/Polypyrrole Nanocomposite as Electrochemical Supercapacitor Electrode: Electrochemical Impedance Studies. *Graphene*, *2*, 81–87. doi:10.4236/graphene.2013.22012

Berhanuddin, N. I. C., Zaman, I., Rozlan, S. A. M., Karim, M. A. A., Manshoor, B., Khalid, A., … Meng, Q. (2017). Enhancement of mechanical properties of epoxy/graphene nanocomposite. *Journal of Physics Conference Series, 914*.

Björk, J., Hanke, F., Palma, C.-A., Samori, P., Cecchini, M., & Persson, M. (2010). Adsorption of Aromatic and Anti-Aromatic Systems on Graphene through π-π Stacking. *The Journal of Physical Chemistry Letters*, *1*(23), 3407–3412. doi:10.1021/jz101360k

Bordes, E., Morcos, B., Bourgogne, D., Andanson, J.-M., Bussière, P.-O., Santini, C. C., … Padua, A. A. H. (2019). Dispersion and Stabilization of Exfoliated Graphene in Ionic Liquids. *Frontiers in Chemistry, 7*(223), 1-14.

Bourlinos, A. B., Georgakilas, V., Zboril, R., Steriotis, T. A., & Stubos, A. K. (2009). Liquid-Phase Exfoliation of Graphite Towards Solubilized Graphenes. *Small, 5*(16), 1841–1845. doi:10.1002mll.200900242 PMID:19408256

Cahill, D. G. (1990). Thermal conductivity measurement from 30 to 750 K: The 3ω method. *The Review of Scientific Instruments, 61*(2), 802–808. doi:10.1063/1.1141498

Chakrabarti, A., Lu, J., Skrabutenas, J. C., Xu, T., Xiao, Z., Maguire, J. A., & Hosmane, N. S. (2011). Conversion of Carbon dioxide to Few-Layer Graphene. *Journal of Materials Chemistry, 21*(26), 9491–9493. doi:10.1039/c1jm11227a

Ciesielski, A., & Samori, P. (2014). Graphene via sonication assisted liquid-phase exfoliation. *Chemical Society Reviews, 43*(1), 381–398. doi:10.1039/C3CS60217F PMID:24002478

Clark, S. (2016). *Graphene Market by Product Type (Mono-layer & Bi-layer Graphene, Few Layer Graphene, Graphene Oxide, and Graphene Nano Platelets) and Application (RFID, Composites, Sensors, Research & Development, Energy Storage, Functional Ink, and Polymer Additives) - Global Opportunity Analysis and Industry Forecast, 2014-2021*. Market Research Report.

Cola, B., Xu, X., & Fisher, T. S. (2007). Increased real contact in thermal interface materials. *Applied Physics Letters, 90*, 093513. doi:10.1063/1.2644018

Coleman, J. N. (2013). Liquid Exfoliation of Defect-Free Graphene. *Accounts of Chemical Research, 46*(1), 14–22. doi:10.1021/ar300009f PMID:22433117

Dhand, V., Rhee, K. Y., Kim, H. J., & Jung, D. H. (2013). A Comprehensive Review of Graphene Nanocomposites: Research Status and Trends. *Journal of Nanomaterials, 763953*, 1–14. doi:10.1155/2013/763953

Dorerk, G. S., Carraro, C., & Maboudian, R. (2010). Single Nanowire Thermal Conductivity Measurements by Raman Thermography. *ACS Nano*. PMID:20731463

Edwards, R. S., & Coleman, K. S. (2013). Graphene synthesis: Relationship to applications. *Nanoscale, 5*(1), 38–51. doi:10.1039/C2NR32629A PMID:23160190

Ghosh, S., Bao, W., Nika, D., Subrina, S., Pokatilov, E., Lau, C., & Balandin, A. (2010). Dimensional crossover of thermal transport in few-layer graphene. *Nature Materials, 9*(7), 555–558. doi:10.1038/nmat2753 PMID:20453845

Guo, H.-L., Wang, X.-F., Qian, Q.-Y., Wang, F.-B., & Xia, X.-H. (2009). A Green Approach to the Synthesis of Graphene Nanosheets. *ACS Nano*, *3*(9), 2653–2659. doi:10.1021/nn900227d PMID:19691285

Hameed, N., Dumée, L. F., Allioux, F.-M., Reghat, M., Church, J. S., Naebe, M., … Fox, B. L. (2018). Graphene based room temperature fexible nanocomposites from permanently cross-linked networks. *Scientific Reports, 8*(2803), 1-8.

Hansson, J., Zandzen, C., Ye, L., & Liu, J. (2016). Review of Current Progress of Thermal Interface Materials for Electronics Thermal Management Applications. *Proceedings of the 16th International Conference on Nanotechnology*. 10.1109/NANO.2016.7751383

Huang, X., Qi, X., Boey, F., & Zhang, H. (2012). Graphene-based composites. *Chemical Society Reviews*, *41*(2), 666–686. doi:10.1039/C1CS15078B PMID:21796314

Hummers, W. S. Jr, & Offeman, R. E. (1958). Preparation of Graphitic Oxide. *Journal of the American Chemical Society*, *80*(6), 1339. doi:10.1021/ja01539a017

Jana, M., Saha, S., Khanra, P., Murmu, N. C., Srivastava, S. K., Kuila, T., & Lee, J. H. (2014). Bio reduction of graphene oxide using drained water from soaked mung beans (Phaseolus aureus L.) and its application as energy storage electrode material. *Materials Science and Engineering B*, *186*, 33–40. doi:10.1016/j.mseb.2014.03.004

Khanam, P. N., AlMaadeed, M. A., Ouederni, M., Harkin-Jones, E., Mayoral, B., Hamilton, A., & Sun, D. (2016). Melt processing and properties of linear low-density polyethylene-graphene nanoplatelet composites. *Vacuum*, *130*, 63–71. doi:10.1016/j.vacuum.2016.04.022

Kim, H., Abdala, A. A., & Macosko, C. W. (2010). Graphene/Polymer Nanocomposites. *Macromolecules*, *43*(16), 6515–6530. doi:10.1021/ma100572e

Kim, K. S., Zhao, Y., Jang, H., Lee, S. Y., Kim, J. M., Kim, K. S., ... Hong, B. H. (2009). Large-scale pattern growth of graphene films for stretchable transparent electrodes. *Nature*, *457*(7230), 706–710. doi:10.1038/nature07719 PMID:19145232

Kuilla, T., Bhadra, S., Yao, D., Kim, N. H., Bose, S., & Lee, J. H. (2010). Recent advances in graphene-based polymer composites. *Progress in Polymer Science*, *35*(11), 1350–1375. doi:10.1016/j.progpolymsci.2010.07.005

Kumar, S., Gautam, C., Mishra, V. K., Chauhan, B. S., Srikrishna, S., Yadav, R. S., ... Rai, S. B. (2019). Fabrication of Graphene Nanoplatelet-Incorporated Porous Hydroxyapatite Composites: Improved Mechanical and in Vivo Imaging Performances for Emerging Biomedical Applications. *ACS Omega*, *4*(4), 7448–7458. doi:10.1021/acsomega.8b03473 PMID:31459841

Lee, C., Wei, X. D., Kysar, J. W., & Hone, J. (2008). Measurement of the Elastic Properties and Intrinsic Strength of Monolayer Graphene. *Science*, *321*(5887), 385–388. doi:10.1126cience.1157996 PMID:18635798

Li, X., Cai, W., An, J., Kim, S., Nah, J., Yang, D., ... Ruoff, R. S. (2009). Large-Area Synthesis of High-Quality and Uniform Graphene Films on Copper Foils. *Science*, *324*(5932), 1312–1314. doi:10.1126cience.1171245 PMID:19423775

Li, X., Magnuson, C. W., Venugopal, A., An, J., Suk, J. W., Han, B., ... Ruoff, R. S. (2010). Magnuson, C. W., Venugopal, A., An,J., Suk, J. W., Han, B., Borysiak, M., Cai, W., Velamakanni, A., Zhu, Y., Fu, L., Vogel, E. M., Voelkl, E., Colombo, L., Ruoff, R. S. "Graphene Films with Large Domain Size by a Two-Step Chemical Vapor Deposition Process. *Nano Letters*, *10*(11), 4328–4334. doi:10.1021/nl101629g PMID:20957985

Li, X., Magnuson, C. W., Venugopal, A., Tromp, R. M., Hannon, J. B., Vogel, E. M., ... Ruoff, R. S. (2011). Large-Area Graphene Single Crystals Grown by Low-Pressure Chemical Vapor Deposition of Methane on Copper. *Journal of the American Chemical Society*, *133*(9), 2816–2819. doi:10.1021/ja109793s PMID:21309560

Lin, L., Zhang, J., Su., H., & Li, J. (2019). Towards super-clean graphene. *Nature Communications, 1912*, 1-7.

Lin, Y.-M., Dimitrakopoulos, C., Jenkins, K. A., Farmer, D. B., Chiu, H.-Y., Grill, A., & Avouris, P. (2010). 100-GHz Transistors from Wafer-Scale Epitaxial Graphene. *Science*, *327*(5966), 662. doi:10.1126cience.1184289 PMID:20133565

Liu, N., Luo, F., Wu, H., Liu, Y., Zhang, C., & Chen, J. (2008). One-Step Ionic-Liquid-Assisted Electrochemical Synthesis of Ionic-Liquid-Functionalized Graphene Sheets Directly from Graphite. *Advanced Functional Materials*, *2008*(18), 1518–1525. doi:10.1002/adfm.200700797

Mack, J. J., Viculis, L. M., Ali, A., Luoh, R., Yang, G., Hahn, H. T., ... Kaner, R. B. (2005). Graphite Nanoplatelet Reinforcement of Electrospun Polyacrylonitrile Nanofibers. *Advanced Materials*, *17*(1), 77–80. doi:10.1002/adma.200400133

Majumdar, P., & Chakrabarti, A. (2017). Graphene and Boron Nitride-Based Nanocomposites with Enhanced Thermal Properties. *International Journal of Nanoparticles and Nanotechnology, 3*(10), 1-6.

Marcano, D. C., Kosynkin, D. V., Berlin, J. M., Sinitskii, A., Sun, Z., Slesarev, A., ... Tour, J. M. (2010). Improved Synthesis of Graphene Oxide. *ACS Nano, 4*(8), 4806–4814. doi:10.1021/nn1006368 PMID:20731455

Marconnet, A. M., Yamamoto, N., Panzer, M. A., Wardle, B., & Goodson, K. E. (2011). L and K. E. Goodson, Thermal Conduction in Aligned Nanotube – Polymer Nanocomposites with High Packing Density. *ACS Nano, 5*(6), 4818–4825. doi:10.1021/nn200847u PMID:21598962

Mishra, A. K., & Ramaparbhu, S. (2011). Functionalized Graphene-Based Nanocomposites for Supercapacitor Application. *The Journal of Physical Chemistry C, 115*(29), 14006–14013. doi:10.1021/jp201673e

Moser, J., Barreiro, A., & Bachtold, A. (2007). Current-induced cleaning of graphene. *Applied Physics Letters, 91*.

Nair, R. R., Blake, P., Grigorenko, A. N., Novoselov, K. S., Booth, T. J., Stauber, T., ... Geim, A. K. (2008). Fine Structure Constant Defines Visual Transparency of Graphene. *Science, 320*(5881), 1308. doi:10.1126cience.1156965 PMID:18388259

Novoselov, K. S., Geim, A. K., Morozov, S. V., Jiang, D., Zhang, Y., Dubonos, S. V., ... Firsov, A. A. (2004). Electric field effect in atomically thin carbon films. *Science, 306*(5696), 666–669. doi:10.1126cience.1102896 PMID:15499015

Novoselov, K. S., Jiang, D., Schedin, F., Booth, T. J., Khotkevich, V. V., Morozov, S. V., & Geim, A. K. (2005). Two-dimensional atomic crystals. *Proceedings of the National Academy of Sciences of the United States of America, 102*(30), 10451–10453. doi:10.1073/pnas.0502848102 PMID:16027370

Papageorgiou, D. G., Kinloch, I. A., & Young, R. J. (2017). Mechanical properties of graphene and graphene-based nanocomposites. *Progress in Materials Science, 90*, 75–127. doi:10.1016/j.pmatsci.2017.07.004

Park, W., Guo, Y., Li, X., Hu, J., Liu, L., Ruan, X., & Chen, Y. (2015). High-Performance Thermal Interface Material Based on Few-Layer Graphene Composite. *The Journal of Physical Chemistry C, 119*(47), 26753–26759. doi:10.1021/acs.jpcc.5b08816

Pei, S., & Cheng, H.-M. (2012). The reduction of graphene oxide. *Carbon, 50*(9), 3210–3228. doi:10.1016/j.carbon.2011.11.010

Pop, E., Varshney, V., & Roy, A. K. (2012). Thermal properties of graphene: Fundamentals and applications. *MRS Bulletin*, *37*(12), 1273–1281. doi:10.1557/mrs.2012.203

Pumera, M. (2013). Electrochemistry of graphene, graphene oxide and other graphenoids: Review. *Electrochemistry Communications*, *36*, 14–18. doi:10.1016/j.elecom.2013.08.028

Qian, W., Hao, R., Hou, Y., Tian, Y., Shen, C., Gao, H., & Liang, X. (2009). Solvothermal-Assisted Exfoliation Process to Produce Graphene with High Yield and High Quality. *Nano Research*, *2*(9), 706–712. doi:10.100712274-009-9074-z

Rao, C. N. R., Sood, A. K., Subrahmanyam, K. S., & Govindaraj, A. (2009). Graphene: The New Two-Dimensional Nanomaterial. *Angewandte Chemie International Edition*, *48*(42), 7752–7777. doi:10.1002/anie.200901678 PMID:19784976

Ravula, S., Baker, S. N., Kamath, G., & Baker, G. A. (2015). 'Ionic liquid-assisted exfoliation and dispersion: Stripping graphene and its two-dimensional layered inorganic counterparts of their inhibitions. *Nanoscale*, *7*(10), 4338–4353. doi:10.1039/C4NR01524J PMID:25689172

Sansoz, F. (2011). *Surface Faceting Dependence of Thermal Transport in Silicon Nanowires* (Vol. 11). Nano Letters.

Seol, J., Jo, I., Moore, A., Lindsay, L., Aitken, Z., Pettes, M., ... Shi, L. (2010). Two-Dimensional Phonon Transport in Supported Graphene. *Science*, *328*(5975), 213–216. doi:10.1126cience.1184014 PMID:20378814

Si, Y., & Samulski, E. T. (2008). Exfoliated Graphene Separated by Platinum Nanoparticles. *Chemistry of Materials*, *20*(21), 6792–6797. doi:10.1021/cm801356a

Singh, D., Murthy, J. Y., & Fisher, T. S. (2011). Mechanism of thermal conductivity reduction in few-layer graphene. *Journal of Applied Physics*, *110*(4), 044317. doi:10.1063/1.3622300

Soudi, A., Dawson, R. D., & Gu, Y. (2011). Quantitative Heat Dissipation Characteristics in Current-Carrying GaN Nanowires Probed by Combining Scanning Thermal Microscopy and Spatially Resolved Raman Spectroscopy. *ACS Nano*, *5*(1), 255–262. doi:10.1021/nn102818s PMID:21155591

Stankovich, S., Piner, R. D., Chen, X., Wu, N., Nguyen, S. B. T., & Ruoff, R. S. (2006). Stable aqueous dispersions of graphitic nanoplatelets via the reduction of exfoliated graphite oxide in the presence of poly(sodium-4-styrenesulfonate). *Journal of Materials Chemistry*, *16*(2), 155–158. doi:10.1039/B512799H

Stoller, M. D., Park, S. J., Zhu, Y. W., An, J. H., & Ruoff, R. S. (2008). Graphene-Based Ultracapacitors. *Nano Letters*, *8*(10), 3498–3502. doi:10.1021/nl802558y PMID:18788793

Sutter, P., Sadowski, J. T., & Sutter, E. (2009). Graphene on Pt(111): Growth and substrate interaction. *Physical Review B: Condensed Matter and Materials Physics*, *80*(24), 245411. doi:10.1103/PhysRevB.80.245411

Syamala, R.-R. (2016). *Computational Analysis of IGBT Package Using Graphene Composites as Thermal Interface Materials* (M.S Thesis). Department of Mechanical Engineering, Northern Illinois University.

Toh, S. Y., Loh, K. S., Kamarudin, S. K., & Daud, W. R. W. (2014). Graphene production via electrochemical reduction of graphene oxide: Synthesis and characterisaion. *Chemical Engineering Journal*, *251*, 422–434. doi:10.1016/j.cej.2014.04.004

Varadaraju, V. (2017). *Development of Borin-Arsenide-Based Composites as Thermal Interface Materials for High Power Generating Electronics Modules* (M.S. Thesis). Department of Mechanical Engineering, Northern Illinois University.

Wang, R., Wu, L., Zhuo, D., Zhang, J., & Zheng, Y. (2018). Fabrication of Polyamide 6 Nanocomposite with Improved Thermal Conductivity and Mechanical Properties via Incorporation of Low Graphene Content. *Industrial & Engineering Chemistry Research*, *57*(32), 10967–11097. doi:10.1021/acs.iecr.8b01070

Yang, L., Weng, W., Yang, J., Zhang, Y., Liang, Y., Luo, X., & Zhu, M. (2018). Highly Conductive Nanocomposite Enabled by an Accordion-like Graphene Network for Flexible Heating Films and Supercapacitors. *ACS Appl. Nano Mater.*, *1*(9), 4781–4787. doi:10.1021/acsanm.8b00997

Yavari, F., Fard, H. R., Pashayi, K., Rafiee, M. A., Zamiri, A., Yu, Z., ... Koratkar, N. (2011). Enhanced Thermal Conductivity in a Nanostructured Phase Change Composite due to Low Concentration Graphene Additives. *The Journal of Physical Chemistry C*, *115*(17), 8753–8758. doi:10.1021/jp200838s

Young, R. J., Liu, M., Kinloch, I. A., Li, S., Zhao, X., Vallés, C., & Papageorgiou, D. G. (2018). The mechanics of reinforcement of polymers by graphene nanoplatelets. *Composites Science and Technology*, *154*, 110–116. doi:10.1016/j.compscitech.2017.11.007

Yu, D., & Dai, L. (2010). Self-Assembled Graphene/Carbon Nanotube Hybrid Films for Supercapacitors. *The Journal of Physical Chemistry Letters*, *12*(2), 467–470. doi:10.1021/jz9003137

Yu, H., Zhang, B., Bulin, C., Li, R., & Xing, R. (2016). High-efficient Synthesis of Graphene Oxide Based on Improved Hummers Method. *Scientific Reports, 6*.

Yu, Q., Lian, J., Siriponglert, S., Li, H., Chen, Y. P., & Pei, S.-S. (2008). Graphene segregated on Ni surfaces and transferred to insulators. *Applied Physics Letters*, *93*(11), 113103. doi:10.1063/1.2982585

Zhang, B., Lee, W. H., Piner, R., Kholmanov, I., Wu, Y., Li, H., ... Ruoff, R. S. (2012). Low-temperature chemical vapor deposition growth of graphene from toluene on electropolished copper foils. *ACS Nano*, *6*(3), 2471–2476. doi:10.1021/nn204827h PMID:22339048

Chapter 4
Preparation and Application of Polymer-Metal Oxide Nanocomposites in Wastewater Treatment:
Challenges and Potentialities

Victor Odhiambo Shikuku
Kaimosi Friends University College, Kenya

Wilfrida N. Nyairo
Kaimosi Friends University College, Kenya

ABSTRACT

The search for efficient and sustainable wastewater treatment technologies is a subject of continuing research. This is due to the emergence of new classes of water contaminants that are recalcitrant to the conventional wastewater treatment technologies and the stringent allowable limits for contaminant levels set by environmental management authorities. The chapter discusses the developments in synthesis methods and application of polymer-metal oxides as emerging facile materials for wastewater treatment. The varying uses of polymer-metal oxides for different processes in water treatment under varying operational conditions and their performance for different pollutants are critically analyzed. Their strengths and inherent limitations are also highlighted. The chapter demonstrates that polymer-metal oxides are facile low-cost and efficient materials and can be integrated in wastewater and drinking water treatment systems.

DOI: 10.4018/978-1-7998-1530-3.ch004

INTRODUCTION

Industrial and other anthropogenic activities have resulted to increased stress in access to clean water. These activities, though intended for beneficial purposes, are associated with release of toxic organic and inorganic chemicals in water systems posing a threat to the well-being of human, animal and aquatic life. In emerging economies mostly characterized with insufficient water treatment facilities and large populations are not connected to the municipally treated water and therefore depend on river and ground water, chemical and biological contamination of water is a serious problem whose solution needs no further delay. Similar trends are reported worldwide (Suriyaraj and Selvakumar, 2016). Indiscriminate disposal of agrochemicals and ad-hoc discharge of industrial effluents laden with chemicals such as dyes, pharmaceutical compounds, detergents, personal care products, the so called chemicals of emerging concern among others are among the leading causes of water pollution. The compounds have been demonstrated to cause of induce development of different illnesses and disorders in both human and aquatic life raising serious environmental concerns (Yang, 2011; Bottoni et al., 2010). These concerns have necessitated development of more stringent drinking water quality parameters by environmental and water resources management authorities. Unfortunately, most the chemical contaminants are recalcitrant to the conventional water treatment techniques such as coagulation and adsorption onto activated carbon only and the compounds find way to drinking and potable water (Shannon et al., 2008). Despite emergent technologies such as membrane technologies and chemical treatment being effecting in sequestering most contaminants from water, they generally invite high capital investment and have high operational costs hence are not widely used especially in developing countries (Jo et al., 2016). The search of low-cost, effective and sustainable technology (materials and processes) for production of potable drinking water is a subject of on-going research among scientists. Among the proposed alternative techniques for water treatment include the advanced oxidation processes (AOPs), such as the Fenton reaction, ozonation, and photocatalysis attributed to their simplicity, relatively low cost, high efficacy and ease of operation (Chong et al., 2010). Noteworthy, each of these techniques possess their inherent limitations. In the recent decades, metal oxide/polymer nanocomposites have been shown to be facile low cost and environmentally friendly materials for water treatment. They have been applied as adsorbents for various pollutants, as photocatalysts for degradation of organic compounds, as membrane filters and for bacteriological treatment of water. This chapter describes the synthesis and characterization techniques for development of metal oxide/polymer nanocomposites and their application for various water treatment processes namely, membrane technology, photocatalysis, biological treatment and adsorption. It is demonstrated that metal oxide/polymer nanocomposites are promising materials for water purification systems.

Synthesis of Polymer-Metal Oxide Composites

Polymer-metal oxide nanocomposites are widely applied in wastewater treatment in areas such as membrane technologies and photocatalytic degradation of pollutants. The commonly employed routes of synthesis of these nanocomposites include; *in situ* chemical oxidative polymerization, electropolymerization, photo-induced polymerization, microemulsion, electrospinning and phase inversion among others.

In situ chemical oxidative polymerization involves dispersal of the metallic oxide nanoparticles in a solvent using ultrasonication vibrations. The monomer is then introduced to the slurry containing the nanoparticles which is thoroughly mixed before initiating polymerization using a chemical oxidant (Karim, Lee and Lee, 2006; Zhu et al, 2008; Teli et al, 2013; Cao et al, 2015; Luan et al 2017). Zhu et al (2008) synthesized a composite of polythiophene/titanium dioxide (PT/TiO_2) for photocatalytic degradation of methyl orange dye using UV light irradiation through the chemical oxidative polymerization route. The TiO_2 nanoparticles were first dispersed in chloroform; thiophene was then added to the resultant dispersion followed by addition of $FeCl_3$ solution to initiate polymerization.

Electro-polymerization involves electrodeposition of conductive polymers onto the metallic oxide nanoparticles (Kickeibick et al, 2006; Ivanovici et al, 2008; Nam et al, 2011). Nam et al (2011) used an elcetropolymerization technique (Atom Transfer radical polymerization technique) for surface modification of titanium dioxide with poly(hydroxylethyl ethacrylate). The electropolymerization route is also reported for the synthesis of polythiophene and titanium dioxide (PTh/TiO_2) composite by Liang and Li (2009)

Photo-induced polymerization involves polymerization of a monomer initiated by metal oxide nanoparticles in a suspension under UV light irradiation (Weng and Ni, 2008; Xu et al, 2011). Weng and Ni (2008) utilized photoexcited TiO_2 to polymerize thiophene which was applied in the photocatalytic degradation of rhodamine B.

Microemulsion involves the formation of an oil (monomer) and water emulsion in the presence of a surfactant and subsequent polymerization (Brijmohan and Shaw 2007; Sun et al 2013). In the synthesis of ɣ-Fe_2O_3 sulfonated polystyrene Brijmohan and Shaw (2007) made an emulsion of hydrophobic ɣ-Fe_2O_3, surfactant and styrene before initiating polymerization using potassium persulfate.

Electrospinning is a fiber production method that uses an electric force to draw charged threads of polymer solutions into fiber (Tan et al, 2017; Zhang et al, 2017). Figure 1 is an illustration of development of polymer-metal oxides fibers using the electrospinning method (Mondal, 2017).

Figure 1. Synthesis of metal oxide polymer composites using electrospinning technique

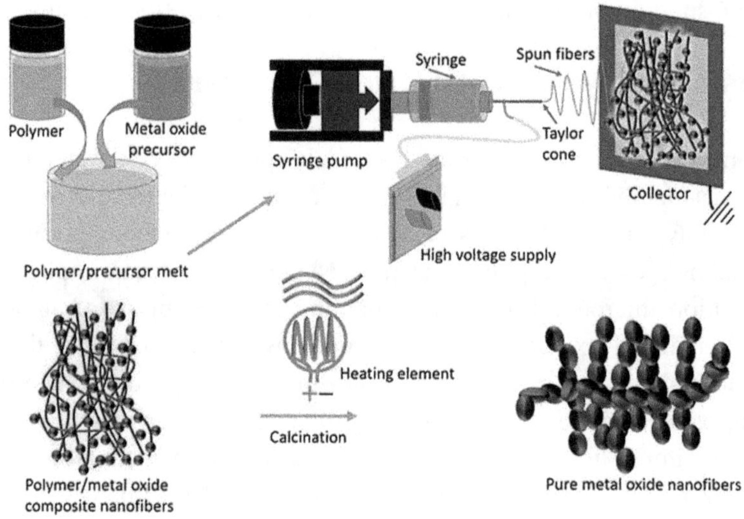

Phase inversion is the process widely used for the preparation of membranes for microfiltration and ultrafiltration (Yu et al, 2015; Krishnamurthy et al, 2016). The process involves transformation of the polymer solution from liquid to solid state. Before solidification liquid-liquid demixing is induced by immersion precipitation (immersion in nonsolvent bath), controlled evaporation or thermal precipitation. Demixing leads to the formation of polymer-rich and polymer-lean phases, the polymer rich phase solidifies first through gelation or crystallization resulting to the formation of the solid membrane. Table 1 shows the route of synthesis some metal oxide–polymer composites

Characterization of Metal-Oxide Nanocomposites

As in any branch of materials science, a detailed characterization of metal oxide/ polymer nanocomposite is crucial. The essential features include topography of surfaces, composition, morphology, crystallinity, shape, and size and standardized instrumentation and techniques. Below is a description of the most common techniques used in characterization of nanocomposites.

Table 1. The methods of synthesis of metal oxide–polymer composites

Nanocomposite	Method of Synthesis	Reference
PTh/TiO_2	Electropolymerization	Liang and Li (2009)
Polypyrrole/polyvinyl alcohol/ TiO_2	*In situ* chemical oxidative polymerization	Cao et al (2015)
PTh/TiO_2	Photo induced polymerization	Weng and Ni (2008)
Polypyrrole/TiO_2	Microemulsion polymerization	Sun et al (2013)
Cu_2O/poly(ethersulfone)	Phase inversion	Krishnamurthy et al (2016)
Polysulfone/TiO_2	Electrospinning	Zhang et al (2017)

Raman Spectroscopy

When light is scattered from a molecule or crystal, photons are mostly elastically scattered. The scattered photons possess the same energy and therefore frequency as the incident photons. However, a limited fraction of the incident light is scattered at optical frequencies different from and less than, the frequency of the incident photons. The process leading to this inelastic scatter is termed the Raman effect. Raman scattering is associated with changes in vibrational, rotational, translation or electronic energy of a molecule. If the scattering is elastic, the process is called Rayleigh scattering. If it's not elastic, the process is called Raman scattering Raman spectroscopy determines the vibrational frequencies of molecules that are Raman active and these frequencies rely on the mass and bond strength of atoms. Raman scattering was discovered by C. V. Raman in 1928 for which he won the Nobel prize.

Fourier-Transform Infra-Red (FT-IR) Spectroscopy

Infrared spectroscopy deals with the interaction of electromagnetic radiation of infrared frequency with molecules. Infrared spectroscopy is one of the most useful analytical tool to perform structural analysis. Provided that the molecule under examination is infrared active, absorbs Infrared radiation, then different types of structural information can be obtained. FT-IR is used to differentiate functional groups present in an unknown compound or material.

X-Ray Diffraction (XRD)

XRD is a rapid and nondestructive analytical technique primarily used for phase identification of crystalline substances and provides information on unit cell dimensions as well. The analyzed material is finely ground, homogenized, and average bulk composition is determined. Additional information acquired through XRD analysis includes crystal structure, microstructure, chemical composition, lattice constants, and particle size of a material.

Scanning Electron Microscopy- Energy Dispersive X-Ray Spectroscopy (SEM-EDS)

SEM-EDS is a method for high-resolution imaging of surfaces and determination of the elemental or chemical composition of a sample. Here, the electrons interact with the atoms constituting the sample, generating signals that contain information about the sample's surface topography, chemical composition, besides other properties such as electrical conductivity.

The method is based on the fundamental principle that each element has a unique atomic structure allowing X-rays that are characteristic of an element's atomic structure to be identified uniquely from one another.

Ultraviolet (UV)-Visible Spectroscopy

This spectrophotometric technique, as the name implies, involves the measurement of light photons within the UV-visible region. The intensity of light before and after passing through the material is measured and the information can be used for both qualitative (structural characteristics) and quantitative analysis.

Nuclear Magnetic Resonance (NMR) Spectroscopy

This technique depends on the population of magnetic nuclei in an external magnetic field to align the nuclei in a finite and predictable number of orientations. NMR gives information on the environment in which the nuclei of atoms are found.

Thermogravimetric Analysis (TGA)

This technique is used to measure the weight of a material as a function of material time or temperature at a constant heat rate. TGA is based on heating a mixture of materials at a high temperature to decompose them in the gas phase. The TGA results are generally obtained as a curve with a percent weight against temperature under controlled atmosphere.

Application of Polymer-Metal Oxide in Water Treatment

Polymer-Metal Oxide in Membrane Technologies

A membrane acts as a selective barrier between two phases. Membranes have recently received more attention for wastewater treatment because they do not require additives and they do not produce harmful by-products (Owen et al, 1995). However, there are challenges associated with membrane technology such as membrane fouling and flux decline during operation (Wang et al, 2000). The polymeric membranes commonly used include: polyacrylonitrile (PAN), polypropylene (PP), polysulfone (PSF), polytetrafluoroethylene (PTFE), polyimide, polyethersulfone (PES) and polyvinylidene fluoride (PVDF) because they are considered to be relatively inexpensive and they are associated with good pore-forming control (Ulbricht, 2006). Conversely, the inorganic membranes composed of metal oxides are highly durable because they are more resistant to chemical attack, extreme pH and oxidation but they are relatively expensive (Labbez et al, 2006). Therefore, metal oxide nanoparticles such as TiO_2 have been introduced to polymers to produce membranes that have synergic properties between the inorganic particles and the polymeric material. These composites and their applications are discussed in the following section.

TiO_2 has been studied for wastewater treatment because it has photocatalytic properties that degrade organic compounds and bacteria (Mills and le Huntes, 1997). TiO_2/polymer composite membranes have been effectively applied in the photodegradation of pollutants in water treatment (Ursino et al, 2018). Teli et al (2013) prepared polysulfone-PANI/TiO_2 by immobilizing TiO_2 onto polyaniline through *in situ* polymerization and the composite showed enhanced properties such as membrane antifouling and reduced agglomeration. Electrospun polysulfone (PSF)/ titanium oxide TiO_2 nanocomposite fibre for forward osmosis membranes synthesized by Zhang et al (2017) showed higher water flux compared to commercial membranes.

A study on the influence of copper oxide (Cu_2O) nanomaterials in a poly(ethersulfone) membrane for improved humic acid and water-oil separation was carried out by Krishnamurthy et al (2016). The composite prepared via phase inversion was found to have a higher water flux rate and desired hydrophilic properties. Yu et al (2015) prepared nano-SiO_2/PVC composite ultrafiltration membrane through phase inversion process. The membrane showed better antifouling performance and higher flux rate compared to bare membranes.

Polymer-Metal Oxides as Photocatalysts

Photocatalytic degradation of pollutants in wastewater is an effective technique of wastewater treatment because it does not require energy input and it does not produce harmful waste. The photocatalytic process involves photoactivation of metal oxide by a photon from the abundant sunlight. When the metal oxide is irradiated electrons and holes are generated if the energy of the incident photons is greater or equal to the metal oxide band gap energy. The electron-hole pairs created initiate photocatalytic reactions (Park et al, 2014). Numerous metal oxide based photocatalysts have been studied in the degradation of highly toxic compounds (Chen et al, 2015). However, the major limitations of these metal oxides is the wide band gap ~3.2 eV or greater which is mainly in the UV regime hence visible light which comprises of most of the solar energy is not absorbed (Patel et al, 2014).This limitation has been overcome by tinkering with the metal oxide nanoparticles in order to lower their band gap through processes such as metal or non-metal doping (Wang et al, 2014), metal oxide/polymer composites (Sun et al 2013; Mukherjee, Barghi and Ray, 2014) among others. Polymeric materials such as polythiophene, polyvinylpyrrolidine, polyacrylonitrile, polyaniline, polypyrrole, polyvinyalcohol have been used to fabricate metal oxide to form metal oxide/polymer hybrids for photocatalytic degradation (Ansari et al, 2015; Cao et al, 2015; Luan et al, 2017). Polypyrrole/polyvinyl alcohol-TiO_2 (PPy/PVA-TiO_2) composite prepared through *in situ* polymerization of sol-TiO_2 in pyrrole/PVA solution was reported for photocatalytic degradation of rhodamine B (Cao et al, 2015). The study indicated that PPy/PVA-TiO_2 films had better photocatalytic properties than TiO_2 films and that the composite's photoactivity under both UV and visible light irradiation had insignificant decrease after four recycle experiments. Liang and Li (2009) synthesized of polythiophene-TiO_2 nanotube films (PTh-TNT) through a two-step electrochemical process of anodization and electropolymerization. The study on the photoactivity of the PTh-TNT composite showed significant degradation of 2,3-dichlorophenols (2,3-DCP) under both UV and visible light irradiation (Liang and Li 2009). Synthesized through *in situ* chemical polymerization, polyaniline/$BiYTi_2O_7$ showed higher activity than $BiYTi_2O_7$ or nitrogen doped TiO_2 for photocatalytic degradation of Azocarmine G under visible light irradiation (Luan et al 2017). Sun et al (2013) conducted a study on the photocatalytic activity of polypyrrole/TiO_2 for the degradation of methyl orange. The composite synthesized via reverse microemulsion polymerization showed a better performance in the photocatalytic degradation of methyl orange under natural light than neat TiO_2. The polythiophene/titanium dioxide composite (PTh/TiO_2) synthesized via photoinduced polymerization of thiophene in TiO_2-chloroform suspension was studied for photocatalytic degradation of rhodamine B (RhB) under UV and visible light irradiation (Xu et al, 2011). The activity under UV irradiation was found to be

76% within 180 min while it was 98% under visible light irradiation in 10 h. The photocatalysis mechanism studies indicated that the degradation of RhB by PTh/TiO_2 under UV irradiation was as a result of the photogenerated holes while under visible light irradiation the radicals were the main oxidant for RhB degradation.

Metal-Oxide Nanocomposites for Biological Treatment

Biological water pollutants emanate from pathogens and free-living microbes, such as protozoa, moulds, viruses, and bacteria (Dugan and Williams, 2006). The pathogenic microorganisms find their way to water systems through indiscriminate sewage discharge or through wastewaters from industries such as slaughterhouses. In humans, viruses and bacteria can cause water borne diseases, such as cholera, typhoid, dysentery and polio among others. Therefore, public health safety in regards to occurrence of pathological water contaminants cannot be overstated. Generally, biological pollutants are categorized as microorganisms, biological toxins, and natural organic matter (Dugan and Williams, 2006). The biological treatment strategies applied for sequestration of contaminants from water are widely used due to the less chemical usage, low cost, and environmentally friendly nature of biological treatments. The biological treatment process entails biodegradation, microbiological or enzymatic decomposition of wastes into less toxic and simpler forms by use of microorganisms such as fungi, bacteria, or algae.

Biological treatment processes can either be aerobic or anaerobic depending on the environmental conditions and the products vary depending on the type of process. In anaerobic treatment, the products include methane, biomass, and carbon dioxide. On the other hand, the products formed under aerobic conditions are water, biomass, and carbon dioxide. Application of biological processes to sequester organic compounds such as dyes from wastewaters has been reported in literature. For instance, Paul et al. (2013) reported significant enhancement of microbial decoloration and decomposition of Reactive Red 120 dye using *Pseudomonas sp.* after pretreatment with low dose irradiation. In a separate Garg and co-workers (2012) reported the biodecolorization of textile dye Orange II dye using *Pseudomonas putida*. A dye removal efficiency of 92% was achieved at pH 8, 303 K temperature in less than 100 minutes. Despite these impressive performances, biological treatment techniques are still incapable of eliminating a wide spectrum of contaminants and their metabolites especially the highly polar and water soluble contaminants (Servos et al., 2005). Since the past decade, there has been reports on the potential of Zn, Ag, and Ti based nanomaterials as disinfectants against certain waterborne disease-causing microbes (Jain and Pradeep, 2005). For example Kumar et al. (2016) prepared a ternary polyaniline/TiO_2/graphene nanocomposite. The as-synthesized posted stupendous results as a photocatalyst with high antibacterial activity toward *Enterobacter ludwigii* and *Escherichia coli* attributed to the low recombination of the graphene electron scavenging property coupled with the sensitizing effect of polyaniline. Elsewhere, Jo and others (2016) reported the antibacterial and hydrophilic characteristics of poly(ether sulfone) composites containing zinc oxide nanoparticles as an ultrafiltration membrane. The filters possessed commendable water flux, antifouling properties and relatively high antibacterial activities. In a separate study, Alaoui et al. (2011) reported bacterial activity of anatase-loaded microporous poly(vinylidene fluoride) membrane. The as-fabricated membrane revealed a strong bactericide effect relative to the membrane coupled with UV light only. These studies cumulatively bespeak of the potential of metal oxide nanocomposites as facile and environmentally friendly materials for biological treatment of water.

Metal Oxide Nanocomposites as Adsorbents

Some of the techniques of water treatment include filtration, desalination, adsorption, osmosis, sedimentation, and disinfection (Shannon et al., 2008). However, of all the aforementioned methods, adsorption remains the most widely used technique due to several inherent advantages over the other techniques such simplicity, cost-effectiveness and insensitivity to toxicity (Dubey et al., 2009). Adsorption is a surface phenomenon where molecules accumulate on the surface of a material, known as adsorbent. Adsorption processes are categorized as chemisorption or physisorption depending on the forces involved in adsorbate-adsorbent interactions (Gupta, 2009). Some of the interaction mechanisms include Coulombic, hydrogen bonding, and van der Wall interactions (Ahman et al., 2015). The cost effectiveness of the adsorption process in terms of capital investment involved against the volume of water treated explains why it is predominant use especially in emerging economies. Isothermal modeling is prerequisite for assessing the distribution of the adsorbates between the aqueous and solid phases and the mechanisms of the adsorbate–adsorbent interactions (Shikuku et al., 2018). The nature of specific adsorption mechanisms can be postulated from the shape of the experimental isotherms. Giles et al. (1974) classified isotherms for adsorption of organic solutes into four main groups: L, S, H, and C and thereafter into subgroups. The description of the sub-groups is beyond the scope of the present chapter. Among the most widely used adsorption isotherms are Langmuir, Freundlich and Temkin isotherms. A brief outline of the isotherms is given below:

Langmuir Isotherm

Langmuir (1918) developed an empirical model that assumed a monolayer adsorption of molecules onto a morphologically homogeneous surface containing at a fixed number of active binding sites, with no lateral molecular interactions. Langmuir postulated that all the adsorption sites are identical in both affinity and energy. The model further assumes that the intermolecular attractive forces diminish steadily with increased distance. The original non-linear form of the Langmuir equation is expressed as:

$$q_e = \frac{Q_0 K_L C_e}{1 + K_L C_e} \tag{1}$$

Where q_e is the amount of solute adsorbed per unit mass of the adsorbent at equilibrium (mg/g), C_e is the residual adsorbate concentration in the solution at equilibrium (mg/L), Q_0 is the theoretical maximum adsorption capacity (mg/g) and b is the Langmuir constant related to the free energy of adsorption (L/g). Equation (1) can be linearized and the five linear forms of the Langmuir equation are presented in Table 2.

Freundlich Isotherm

The Freundlich model (Freundlich, 1906) envisions a multilayer adsorption process, with non – uniform distribution of adsorption enthalpy and affinities onto the heterogeneous adsorbent surface but likewise without lateral interactions. The energetically favored binding sites are postulated to be occupied first and the binding strength decreases sequentially with increased coverage of the adsorption sites. The non-linear Freundlich equation is expressed as:

$$q_e = K_F C_e^{1n} \tag{2}$$

The exponential relation indicates that the amount of adsorbate migrating onto the adsorbent surfaces increases with rise in solute concentration. According to Halsey (1952), the Freundlich maximum adsorption capacity is given by the equation:

$$K_F = \frac{q_m}{C_i^{1n}} \tag{3}$$

Table 2. Linear Isotherm Model parameters

Isotherm Model	Equation	Parameters	Plot
Langmuir-1	$\frac{1}{q_e} = \frac{1}{Q_o} + \frac{1}{Q_o K_L C_e}$	Q_o (mg g^{-1}) K_L (L g^{-1})	$\frac{1}{q_e}$ vs $\frac{1}{C_e}$
Langmuir-2	$\frac{C_e}{q_e} = \frac{C_e}{Q_o} + \frac{1}{Q_o K_L}$	Q_o (mg g^{-1}) K_L (L g^{-1})	$\frac{C_e}{q_e}$ vs C_e
Langmuir-3	$q_e = -\frac{q_e}{K_L C_e} + Q_o$	Q_o (mg g^{-1}) K_L (L g^{-1})	q_e vs $\frac{q_e}{C_e}$
Langmuir-4	$\frac{q_e}{C_e} = -K_L q_e + K_L Q_o$	Q_o (mg g^{-1}) K_L (L g^{-1})	$\frac{q_e}{C_e}$ vs q_e
Langmuir-5	$\frac{1}{C_e} = \frac{K_L Q_o}{q_e} - K_L$	Q_o (mg g^{-1}) K_L (L g^{-1})	$\frac{1}{C_e}$ vs $\frac{1}{q_e}$
Freudlich	$\ln q_e = \ln K_f + \frac{1}{n} \ln C_e$	K_f, n	$\ln q_e$ vs $\ln C_e$
Temkin	$q_e = B \ln A_T + B \ln C_e$	A_T, B	q_e vs $\ln C_e$

Where C_i is the initial concentration of the solute in the bulk solution (mg/L) and q_m is the Freundlich maximum adsorption capacity (mg/g). The linearized Freundlich equation is given in Table 2.

Temkin Isotherm

The Temkin isotherm model (Temkin, 1940) assumes that the heat of adsorption of all the molecules in the layer reduces linearly rather than logarithmically with surface coverage due to adsorbent–adsorbate interactions. The adsorption is considered a uniform distribution of the binding energies, up to some maximum binding energy. The non-linear Temkin isotherm equation is given by:

$$q_e = \frac{RT}{b_T} \ln\left(A_T C_e\right) = B \ln\left(A_T C_e\right) \tag{4}$$

Where T is temperature (K), R is the universal gas constant (8.314 J/mol.K), A_T is the equilibrium binding constant (L/mg), b_T represents the variation in adsorption energy (kJ/mol) and B is Temkin constant associated with the parameter b_T by the relation:

$$b_T = \frac{RT}{B} \tag{5}$$

The linear form of Temkin isotherm is given in Table 2.

Adsorption rates are a function of environmental factors such as adsorbate concentration, solution pH and temperature as well as chemical characteristics of the adsorbing molecules and the nature of the adsorbing surfaces. Beside other textural characteristics, a suitable adsorbent must have relatively high surface area. This is achievable by reducing the particle sizes of the adsorbents to nano-scale (Gehrke et al., 2015). A broad spectrum of materials has been evaluated as adsorbents for various pollutants. These include zeolites, biochars, clays and geopolymers among others (Shikuku et al., 2015; Shikuku et al., 2016; Shikuku et al., 2018; Shikuku et al., 2019). Metal oxide polymer composites have been reported to have comparatively high adsorption capabilities as the traditional adsorbents. For example, polypyrrole, polyfuran, and polyethyleneimine exhibit a high affinity for metal ions attributed to the coulombic attractions between the cations ions of the metal oxide and the lone pair of the polymers as demonstrated by Abdi et al. (2009). Mahmud et al. (2014) reported the adsorption of nickel ions onto polypyrrole conducting polymer. The adsorption potential of the polymer was attributed to the presence of positively charged nitrogen atoms in polypyrrole that imbues the material with adsorptive surfaces.

CONCLUSION

This chapter presents a short review and basics on the synthesis, characterization and application of inorganic metal oxide polymer nanocomposites as potential materials for water treatment. The potentialities of polymer/metal oxide nanocomposites as photocatalysts, adsorbents, biological treatment and membrane filtration have been discussed. Polymer nanocomposites possess relatively high mechanical strength, long-term stability, and low-cost synthesis processes while the metal oxides have excellent electronic, magnetic, and catalytic properties. Consequently, the metal oxide/polymer matrices exhibit excellent properties as a result of the synergism of the properties of the building blocks imbuing them with not only adsorptive capabilities but also antibacterial and photocatalytic activities. Besides their excellent performance, they suffer several challenges such as relatively high operational costs, technical skill is required, mechanisms for recovery and regeneration of the exhausted material, varying photostability and environmental concerns associated with handling nanoparticles. The results show that metal oxide polymer nanocomposites should not be used singly but integrated with other water treatment techniques. Theoretical and empirical data are needed for optimization and integration of metal oxide polymer nanocomposites in water treatment.

REFERENCES

Abdi, M. M., Kassim, A., Mahmud, H. E., Yunus, W. M. M., Talib, Z. A., & Sadrolhosseini, A. R. (2009). Physical, optical, and electrical properties of a new conducting polymer. *Journal of Materials Science*, *44*(14), 3682–3686. doi:10.100710853-009-3491-y

Ahmad, A., Mohd-Setapar, S. H., Chuong, C. S., Khatoon, A., Wani, W. A., Kumar, R., & Rafatullah, M. (2015). Recent advances in new generation dye removal technologies: Novel search for approaches to reprocess wastewater. *RSC Advances*, *5*(39), 30801–30818. doi:10.1039/C4RA16959J

Alaoui, O. T., Nguyen, Q. T., Schaetzel, P., & Mbareck, C. (2011). Dye and bacteria photodegradations with anatase-loaded microporous poly (vinylidene fluoride) membranes. *Catalysis Science & Technology*, *1*(8), 1412–1422. doi:10.1039/c1cy00179e

Ansari, M. O., Khan, M. M., Ansar, S. A., & Cho, M. H. (2015). Polythiophene nanocomposites for photodegradation applications: Past, present and future. *Journal of Saudi Chemical Society*, *9*(5), 494–504. doi:10.1016/j.jscs.2015.06.004

Bottoni, P., Caroli, S., & Caracciolo, A. B. (2010). Pharmaceuticals as priority water contaminants. *Toxicological and Environmental Chemistry*, *92*(3), 549–565. doi:10.1080/02772241003614320

Brijmohan, S. B., & Shaw, M. T. (2007). Magnetic ion-exchange nanoparticles and their application in proton exchange membranes. *Journal of Membrane Science*, *303*(1-2), 64–71. doi:10.1016/j.memsci.2007.06.066

Cao, S., Zhang, H., Song, Y., Zhang, J., Yang, H., Jiang, L., & Dan, Y. (2015). Investigation of polypyrrole/polyvinyl alcohol/titatium oxide composite films for photocatalytic applications. *Applied Surface Science*, *342*, 55–63. doi:10.1016/j.apsusc.2015.02.139

Chen, J., Riu, F., Xu, W., Cao, S., & Zhu, H. (2015). Recent progress in enhancing photocatalytic efficiency of TiO_2 –based materials. *Applied Catalysis A, General*, *495*, 131–140. doi:10.1016/j.apcata.2015.02.013

Chong, M. N., Jin, B., Chow, C. W., & Saint, C. (2010). Recent developments in photocatalytic water treatment technology: A review. *Water Research*, *44*(10), 2997–3027. doi:10.1016/j.watres.2010.02.039 PMID:20378145

Dugan, N. R., & Williams, D. J. (2006). Cyanobacteria passage through drinking water filters during perturbation episodes as a function of cell morphology, coagulant and initial filter loading rate. *Harmful Algae*, *5*(1), 26–35. doi:10.1016/j.hal.2005.04.001

Garg, S. K., Tripathi, M., Singh, S. K., & Tiwari, J. K. (2012). Biodecolorization of textile dye effluent by Pseudomonas putida SKG-1 (MTCC 10510) under the conditions optimized for monoazo dye orange II color removal in simulated minimal salt medium. *International Biodeterioration & Biodegradation*, *74*, 24–35. doi:10.1016/j.ibiod.2012.07.007

Gehrke, I., Geiser, A., & Somborn-Schulz, A. (2015). Innovations in nanotechnology for water treatment. *Nanotechnology, Science and Applications*, *8*, 1–17. doi:10.2147/NSA.S43773 PMID:25609931

Gilles, C., MacEwan, T., Nakhwa, S., & Smith, D. (1960). A system of classification of solution adsorption isotherms, and its use in diagnosis of adsorption mechanisms and measurement of specific surface areas of solids. *Journal of the Chemical Society*, *2*, 3973–3993. doi:10.1039/jr9600003973

Gupta, V., & Suhas. (2009). Application of low-cost adsorbents for dye removal—A review. *Journal of Environmental Management*, *90*(8), 2313–2342. doi:10.1016/j.jenvman.2008.11.017 PMID:19264388

Ivanovici, S., Peterlik, H., Fieldgitscher, C., Puchberger, M., & Kickelbick, G. (2008). Atom transfer radical polymerization of complexes bases on Ti and Zr alkoxides modified with β-keto ester ligands and transformation of the resulting polymers in nanocomposites. *Micromolecules*, *41*(4), 1131–1139. doi:10.1021/ma0710838

Jain, P., & Pradeep, T. (2005). Potential of silver nanoparticle-coated polyurethane foam as an antibacterial water filter. *Biotechnology and Bioengineering*, *90*(1), 59–63. doi:10.1002/bit.20368 PMID:15723325

Jo, Y. J., Choi, E. Y., Choi, N. W., & Kim, C. K. (2016). Antibacterial and hydrophilic characteristics of poly(ether sulfone) composite membranes containing zinc oxide nanoparticles grafted with hydrophilic polymers. *Industrial & Engineering Chemistry Research*, *55*(28), 7801–7809. doi:10.1021/acs.iecr.6b01510

Karim, M. R., Lee, C. J., & Lee, M. S. (2006). Synthesis and characterization of conducting polythiophene/carbon nanotubes composites. *Journal of Polymer Science. Part A, Polymer Chemistry*, *44*(18), 5283–5290. doi:10.1002/pola.21640

Kickelbick, G., Holzinger, D., Rutzinger, D., & Ivanovici, S. (2006). Synthesis of inorganic- polymer nanocomposites by atom transfer radical polymerization using initiators or polymerization groups attached to transition metals via bidentate ligands. ACS Symposium Series, 944, 269-278.

Krishnamurthy, P. H., Yogarathinam, L. T., Gangasalam, A., & Ismail, A. F. (2016). Influence of copper oxide nanomaterials in a poly(ethersulfone) membrane for improved humic acid and oil-water separation. *Journal of Applied Polymer Science*, *133*(36), 1–10. doi:10.1002/app.43873

Kumar, R., Ansari, M. O., Parveen, N., Oves, M., Barakat, M., Alshahri, A., ... Cho, M. H. (2016). Facile route to a conducting ternary polyaniline@TiO2/GN nanocomposite for environmentally benign applications: Photocatalytic degradation of pollutants and biological activity. *RSCA Advances*, *6*(112), 111308–111317. doi:10.1039/C6RA24079H

Kumar, V., Talreja, N., Deva, D., Sankararamakrishnan, N., Sharma, A., & Verma, N. (2011). Development of bi-metal doped micro-and nano multi-functional polymeric adsorbents for the removal of fluoride and arsenic (V) from wastewater. *Desalination*, *282*, 27–38. doi:10.1016/j.desal.2011.05.013

Kwak, S.-Y., Kim, S. H., & Kim, S. S. (2001). Hybrid organic/inorganic reverse osmosis (RO) membrane for bactericidal anti-fouling: Preparation and characterization of TiO2 nanoparticle self-assembled aromatic polyamide thin-film-composite (TFC) membrane. *Environmental Science & Technology*, *35*(11), 2388–2394. doi:10.1021/es0017099 PMID:11414050

Labbez, C., Fievet, P., Szymczyk, A., Vidonne, A., Foissy, A., & Pagetti, J. (2002). Analysis of the salt retention of titania membrane using the "DSPM"model: Effect of pH, salt concentration and nature. *Journal of Membrane Science*, *208*(1-2), 315–329. doi:10.1016/S0376-7388(02)00310-1

Liang, H. C., & Li, X. Z. (2009). Visible induced photocatalytic reactivity of polymer sensitized titania nanotube films. *Applied Catalysis B: Environmental*, *86*(1-2), 8–17. doi:10.1016/j.apcatb.2008.07.015

Luan, J., Shen, Y., Wang, S., & Guo, N. (2017). Synthesis, property, characterization and photocatalytic activity of polyaniline/$BiYTi_2O_7$ polymer composite. *Polymers*, *9*(12), 69. doi:10.3390/polym9030069 PMID:30970757

Mahmud, H., Hosseini, S., & Yahya, R. B. (2016). Removal of nickel ions from aqueous solution by polypyrrole conducting polymer. *Key Engineering Materials*, *594*, 793–797.

Mahmud, H. N. M. E., Huq, A. O., & Binti Yahya, R. (2016). The removal of heavy metal ions from wastewater/aqueous solution using polypyrrole-based adsorbents: A review. *RSC Advances*, *6*(18), 14778–14791. doi:10.1039/C5RA24358K

Mills, A., & Le Hunte, S. (1997). An overview of semiconductor photocatalysis. *Journal of Photochemistry and Photobiology A Chemistry*, *108*(1), 1–35. doi:10.1016/S1010-6030(97)00118-4

Mondal, K. (2017). Recent Advances in the Synthesis of Metal Oxide Nanofibers and Their Environmental Remediation Applications. *Inventions*, *2*(2), 1–29. doi:10.3390/inventions2020009

Mukherjee, D., Barghi, S., & Ray, A. K. (2014). Preparation and characterization of the TiO_2 immobilized polymeric photocatalyst for degradation of aspirin under UV and solar light. *Processes*, *2*(1), 12–23. doi:10.3390/pr2010012

Nam, K., Tsutsumi, Y., Yushikawa, C., Tanaka, Y., Fukana, R., Kimura, T., ... Kishida, A. (2011). Preparation of novel polymer-metal oxide nanocomposite with nanophase separation hierarchical structure. *Bulletin of Materials Science*, *34*(7), 1239–1296. doi:10.100712034-011-0317-8

Park, E. J., Jeong, B., Jeong, M. G., & Kim, Y. D. (2014). Synergetic effects of hydrophilic surface modification and N-doping for visible light response on photocatalytic activity of TiO_2. *Current Applied Physics*, *14*(3), 300–305. doi:10.1016/j.cap.2013.12.004

Patel, N., Jaiswal, R., Warang, T., Scarduell, G., Dashara, A., Ahuja, B., ... Miotello, A. (2014). Efficient photocatalytic degradation of organic water pollutants using V-N-doped TiO_2 thin films. *Applied Catalysis B: Environmental*, *150-151*, 74–81. doi:10.1016/j.apcatb.2013.11.033

Paul, J., Kadam, A. A., Govindwar, S. P., Kumar, P., & Varshney, L. (2013). An insight into the influence of low dose irradiation pretreatment on the microbial decolouration and degradation of reactive red-120 dye. *Chemosphere*, *90*(4), 1348–1358. doi:10.1016/j.chemosphere.2012.07.049 PMID:22980957

Servos, M., Bennie, D., Burnison, B., Jurkovic, A., McInnis, R., Neheli, T., ... Ternes, T. A. (2005). Distribution of estrogens, 17β-estradiol and estrone, in Canadian municipal wastewater treatment plants. *The Science of the Total Environment*, *336*(1-3), 155–170. doi:10.1016/j.scitotenv.2004.05.025 PMID:15589256

Shannon, M. A., Bohn, P. W., Elimelech, M., Georgiadis, J. G., Marinas, B. J., & Mayes, A. M. (2008). Science and technology for water purification in the coming decades. *Nature*, *452*(7185), 301–310. doi:10.1038/nature06599 PMID:18354474

Shikuku, V. O., & Tome, S. (2019). Application of Geopolymer Composites in Wastewater Treatment; Trends, Opportunities and Challenges. In Polymer Nanocomposites for Advanced Engineering and Military Applications (pp. 131-149). IGI Global. DOI: doi:10.4018/978-1-5225-7838-3

Shikuku, V. O., & Filipe, F. (2015). A comparison of adsorption equilibrium, kinetics and thermodynamics of aqueous phase clomazone between Faujasite X and a Natural zeolite from Kenya. *South African Journal of Chemistry. Suid-Afrikaanse Tydskrif vir Chemie*, *68*, 245–252. doi:10.17159/0379-4350/2015/v68a33

Shikuku, V. O., & Winfida, N. (2018). Preparation and Application of Biochars for Organic and Microbial Control in Wastewater Treatment Regimes. In Advanced Treatment Techniques for Industrial Wastewater (pp. 19-34). IGI Global. DOI: doi:10.4018/978-1-5225-5754-8.ch002

Shikuku, V. O., Zanella, R., Kowenje, C. O., & Filipe, F. (2018). Single and Binary Adsorption of sulphonamide antibiotics onto iron-modified clay: Linear and nonlinear Isotherms, Kinetics, thermodynamics and mechanistic studies. *Applied Water Science*, *8*(6), 175. doi:10.100713201-018-0825-4

Sun, L., Shi, Y., Li, B., Li, X., & Wang, Y. (2013). Preparation and characterization of polypyrrole/TiO_2 nanocomposites by reverse microemulsion polymerization and its photocatalytic activity for degradation of methyl orange under natural light. *Polymer Composites*, *34*(7), 1076–1080. doi:10.1002/pc.22515

Suriyaraj, S., & Selvakumar, R. (2016). Advances in nanomaterial based approaches for enhanced fluoride and nitrate removal from contaminated water. *RSC Advances*, *6*(13), 10565–10583. doi:10.1039/C5RA24789F

Tan, J. Z., Nursam, N. M., Xia, F., Truong, Y. B., Kyratzis, I. L., Wang, X., & Caruso, R. A. (2017). Electrospun PVDF-TiO_2 with tuneable TiO_2 crystal phases: Synthesis and application In photocatalytic redox reactions. *Journal of Materials Chemistry. A, Materials for Energy and Sustainability*, *5*(2), 641–648. doi:10.1039/C6TA08266A

Teli, S. B., Molina, S., Sotto, A., Calvo, E. G. A., & Abajob, J. D. (2013). Fouling resistant poly-sulfone –PANI/ TiO_2 ultrafiltration nanocomposite membranes. *Industrial and Engineering Research*, *52*(27), 9470–9479. doi:10.1021/ie401037n

Ulbricht, M. (2006). Advanced functional polymer membranes. *Polymer*, *47*(7), 2217–2262. doi:10.1016/j.polymer.2006.01.084

Ursine, C., Castro-Munoz, R., Drioli, E., Gzara, L., Albeirutty, M. H., & Figoli, A. (2018). Progress of nanocomposite membranes for water treatment. *Membranes*, *8*(2), 18. doi:10.3390/membranes8020018 PMID:29614045

Wang, J., Kim, J.-H., Choo, K.-H., Lee, Y.-S., & Lee, C.-H. (2000). Hydrophilic modification of polypropylene microfiltration membranes by ozone-induced graft polymerization. *Journal of Membrane Science*, *169*(2), 269–276. doi:10.1016/S0376-7388(99)00345-2

Wang, Y., Zhang, R., Li, J., Li, L., & Lin, S. (2014). First principles study on transition metal doped anatase TiO_2. *Nanoscale Research Letters*, *9*(1), 46–53. doi:10.1186/1556-276X-9-46 PMID:24472374

Weng, Z., & Ni, X. Y. (2008). Oxidative polymerization of pyrrole photocatalyzed by TiO_2 nanoparticles and interactions in the composites. *Applied Polymer Science*, *110*, 109.

Xu, S., Jiang, L., Yang, H., Song, Y., & Dan, Y. (2011). Structure and photocatalytic activity of PTh/TiO_2 composite particles prepared by photoinduced polymerization. *Chinese Journal of Catalysis*, *32*(3-4), 536–545. doi:10.1016/S1872-2067(10)60207-0

Yang, M. (2011). A current global view of environmental and occupational cancers. *Journal of Environmental Science and Health*, *29*(Part C), 223–249. doi:10.1080/10590501.2011.601848 PMID:21929381

Yu, Z., Liu, X., Zhao, F., Liang, X., & Tian, Y. (2015). Fabrication of a low cost nano-SiO_2/PVC composite ultrafiltration membrane and its antifouling performance. *Journal of Applied Polymer Science*, *132*(2), 1–11. doi:10.1002/app.41267 PMID:25866416

Zhang, C., Huang, M., Meng, L., Li, B., & Cai, T. (2017). Electrospun polysulfone (PSF)/titanium oxide (TiO_2) nanocomposite fibres as subtrates to prepare thin film forward osmosis membranes. *Journal of Chemical Technology and Biotechnology (Oxford, Oxfordshire)*, *92*(8), 2090–2097. doi:10.1002/jctb.5204

Zhu, Y., Xu, S., Jiang, L., Pang, K., & Dan, Y. (2008). Synthesis and characterization of polythiophene/titanium dioxide composite. *Reactive & Functional Polymers*, *68*(10), 1492–1498. doi:10.1016/j.reactfunctpolym.2008.07.008

KEY TERMS AND DEFINITIONS

Adsorption: Accumulation of molecules (adsorbate) from aqueous phase onto a solid material (adsorbent).

Composite: A synthetic material developed by chemical amalgamation of two or more materials with superior properties relative to the individual constituents.

Chapter 5
Nanocomposites in the Food Packaging Industry:
Recent Trends and Applications

Dheeraj Kumar
National Institute of Technology, Durgapur, India

Md. Farrukh
Echelon Institute of Technology, India

Nadeem Faisal
ITM University, Gwalior, India

ABSTRACT

The recent innovations in nanomaterials for the food packaging industry over the conventional food packaging material have made for a better quality of food product. The use of biodegradable materials is environmentally friendly and suitable for maintaining the quality of food. The chapter focuses on nano-composite materials that enhance the antimicrobial, mechanical, thermal, as well as barrier properties against the migrating element in the food packaging system. Bio-composite derivatives such as PLA, PCL, starch and cellulose, protein derivatives of nanocomposite materials have also been discussed in the chapter along with nano-sensors. Aspects of safety for human and environments and need for regulations of hazard assessment for safety purpose for the food packaging have also been discussed in this chapter. The chapter concludes by discussing the use of nanomaterials applications in food packaging for developing countries, forming some conclusions and leaving readers with thoughts for future research directions.

DOI: 10.4018/978-1-7998-1530-3.ch005

INTRODUCTION TO FOOD PACKAGING

Packing has been an essential aspect for the human being for a thousand years. When people started going from one place to another place, they felt the need for packing of the food products. The packaging concept lacked almost a hundred years ago, and food packaging industries were quick to realize this and took the opportunity to fill the gap. Nowadays, the packaging is an essential need of the society, because it encompasses, and protects the goods. The importance of the packaging concept does not need to be justified, or it can be said that its values hardly needs stressing. The reason behind it is that no one can think of selling a product of food items without its packaging. However, knowing all about the importance of packaging and its role, it is quite common to observe, that people still have the mindset that the cost of the packing material is unnecessary and often the price of a product which is high just because of its packaging is taken as a negative aspect by the society. This is only due to the gap or lack of information about what a packaging performs. The people are not known for this information, or maybe because of misunderstanding and lack of information, consumers mostly concentrate on the end-product rather than the packaging of materials. At an earlier time, people were using skins, leaves, and bark materials for the packaging purpose of the food products (Driscoll rh et al., 1999).

The packaging concept holds a significant position in the food processing unit. Now in the present time, much progress, developments have been made in the food packaging industry. It can also be said that in the last three decades, the concept of packaging has been increased in a large volume and is also quite diversified (Coles R et al., 2003). For extending the life of the food product, i.e., Shelf-life, too much innovative idea has been applied in the packaging materials. The packaging is a socio, logical, and scientific discipline of thought which guarantees delivery of products to a consumer who needs those merchandise in the best condition suitable for their utilization. This concept of packing includes a package of food products in the form of pouches, bags, cups, trays, cans, tubes, bottles, or it may contain any container to perform some specific task function to protect the food products. From a survey, which was conducted in the United States, it has been shown that approximately 72% peoples are there in the United States of America who can pay extra money for the freshness of the food products, which have the assured certification of healthy food products and will not harm anymore. So, the concept of food packaging is growing hugely, and a lot of research work is going on for its development and providing better shelf-life of food products for the satisfaction of consumers.

PACKAGING AND PRESERVATION

It is a well-known fact that drying and freezing are the direct approaching techniques for the preservation of foods. Some other methods are there, which are quite necessary to be implemented for the packaging and preservation of food products. The indirect methods are also crucial factors to avoid the phenomenon of contamination or recontamination. These indirect tools are packaging concept and quality management which needs implementation.

Nevertheless, these techniques do not come in the category of food preservation techniques (Rahman et al., 1999). They are giving important consideration for maintaining the quality of the foods most securely and are also quite healthy. There are mainly five functions of the packaging concept. They are also known as (5Ps): and collectively called product containment, preservation, and quality, presentation and convenience, protection, and provide storage history (Figure 1).

Product Containment

The capability of containment and its protection is the first and most important function of the packaging concept. It can be easily explained the reason why the liquids, semiliquid, and powders, etc. like products cannot be marketed without any packaged products. So, for packaging, it needs some containers of some desired quality according to the product kept within the package. So, it results that all the food products must be contained before storing and moving from one place to some other place. Containment talks only about the concept of holding product items compatible with transport also, and protection refers to the prevention against the deterioration of food products (Miltz et al., 1992).

Preservation by Maintaining Quality

The second most important function of the packaging is all about the maintenance of the food product and the local surroundings of its environment. The main objective is to provide enhanced storage life and safety issues. Three factors govern the products shelf-life; they are characteristics of the foods, their properties, and storage with the conditions of distribution for any individual packaging. The reactions which are going inside the packaging system deteriorate the quality of the food products inside the package. Moreover, these reactions include the following: chemical, physical, enzymatic, micro-biological changes occurred. Some other reasons may be due to the presence of insects, pesticides, and rodents also.

Figure 1. 5Ps of Food Packaging

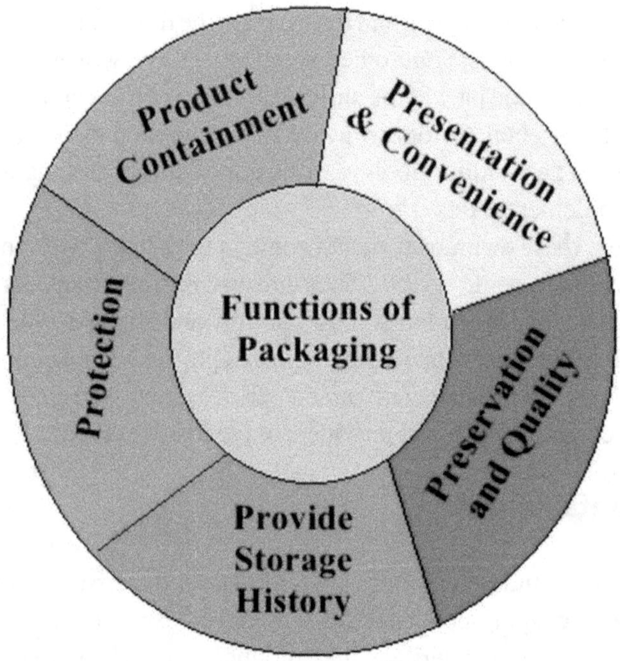

Presentation and Convenience

The third one conceptual function is about the presentation and its convenience. The labels which are available on the food items must have to follow the rules of the organizations of the food industry. These labels give essential information about the products so that the customers who are willing to buy that product can make it. However, customers need some attractiveness as well as the assurance of that product regarding health. A food package can be turned into a compelling package as it needs the presence of well-conditioned and its publicity of the product materials.

Convenience

For a food product, it should fulfill the need of the current, as well as that of future meeting style, and is demanding as per the society. Moreover, some other parameters decide the conveniences of the product. Those are opening, closing, enclosable, tamper-proof, smaller portions. The important thing is about the adaptation of the package size as per the needs of the consumer (size of the family, individuals, special sizes for the food delivery service) are important steps taken by the industries to enhance the conveniences of the food product with the help of packaging. Self-heating containers have already been developed only to provide convenience to the customers. There is no need to heat that food product before consumption again, so it provides convenience in this manner.

Protection during Processing and Distribution

The fourth important function in the food packaging is about protecting at the time when the product is transiting to the customer. Packing is an essential part of the distribution process. So, when the product is going to be delivered to the concern customers, at that time its main objective is to provide the facilities like handling and transportation. It is only the packing which can accept the challenges like heating, humidity or dew, etc. So, there is a need to be aware of the challenges that occurred while distribution, and how to design a comfortable package that fits suitable for that food product. A particular type of frame is there, which provides a base for the carrying purpose of packets.

Provide Storage History

An important consideration for the prediction of microbial concentration and other related parameters responsible for the quality of food products known as Time-temperature indicator (TTI). TTI has been divided into three parts according to the responsive mechanism. They are as follows: (i) Biological (ii) chemical (iii) physical systems. Two key issues are considered during the distribution chain. The first one is economics, and the second is about knowing food products. It should have a better understanding of the degradation kinetics of the products. It will elaborate on the things that how the quality characteristics of the product change and, accordingly the behavior related to time-temperature exposure.

IDEAL PACKAGING

There are no such conditions that exist which represent the ideal packaging. However, it can be tried to come closer to the conditions related to the ideal packaging. So, these are the criteria for an ideal packaging. They are as follows:

- Zero levels of toxicity behavior
- Higher visibility of the food product
- Marketing appeal should be powerful
- Ability to control the moisture and gas content
- Performance should be stable even at a high-temperature zone of working
- Readily available and having low-cost material
- Easy handling machines and low friction coefficient
- It should be properly labeled
- Migration resistant package
- Controlled transmission of the unwanted gases and particles
- It should be protected from the loss of flavor and odor also
- Having suitable mechanical strength properties
- Advanced closure characteristics like; opening, sealing, pouring, and resealing

TYPES OF PACKAGING MATERIAL USED FOR FOODS

At the earlier stage, the materials used for packaging were usually skins, leaves. And barks. However, nowadays, in the industry of food packaging, it has been observed that regarding the materials selections for the food packaging, a lot of developments and research work has been made in order to improve the quality of packing materials for foods. The main objective for materials selection is only to provide diversified materials and equipment, also having desired properties as per the requirement. For the classification of materials for packaging purposes, it has been divided into two parts; rigid and another of flexible type. There are some examples like plastic films, foil, textiles, papers come into the category of flexible nature and woods, glasses, metals, and plastic of hard materials are examples of rigid type packing. For the selection of proper material type as per the requirement, it is vital to consider the properties which needed as a barrier, mechanical, chemical, physical, optical, transport properties, and all.

METALS

For food packaging, the use of metal containers is considered as the most suitable materials. These materials provide almost all the desired properties which should be made available for food packaging. Those properties are mechanically strength behavior, barrier property, impermeable towards mass transfer and also to light, having good conductivity of heat, having the capability to resist high temperature, etc. Considering its opacity property, it is suitable for food products having light sensitivity, but on the other side, the contents inside the package are not visible. However, there are some more disadvantages to these metals can containers. It costs too much high as compared to other materials; mass is also dense, tending to react with the food products inside the metal container and also with an environmental condition resulting like corrosion in it is inside and outside surface.

Under the category of metals, mainly *Tinplate. Steel, Aluminum* is the most useful cans and canister used. The central concept of using metals as a packaging material is to ensure that the product which is kept inside the metal can is stable and seal given to it is complete. One thing significant for the manufacturing of the cans is coating or lacquering materials used. The lacquer is a type of resin material. Some of the examples of it are; acrylic materials having more resistive power towards to sustain in the high-temperature zones, phenolic, polybutadiene, oleo resinous, alkyd, and vinyl resins. In the present age of technologies used for coating materials, there are almost 200 different kinds of protective coatings available to use. If we talk about the thickness of the tin coatings, then its units are expressed in terms of pounds per base box (lb/bb). Food packaging book. The technique was used before 50 years back for the coating of steel plate surface with the tin material was "hot dip" method, and nowadays it has been changed with the technique of electrolytic deposition method. The main advantage of using this technique is that it provides a more uniform tin coating surface with less consumption per unit area. The disadvantage of using the tin coating technique is that if any small gaps are present there, then it will lead to the formation of the corrosion phenomenon.

Furthermore, after corrosion, it produces gases like hydrogen, which can make a possibility of a blow of can. The size of the can is being standardized specified by organizations. And criteria for the selection of cans are steel bae specification, a thin layer thickness of the tin coating, and type of channels, also the geometry of cans.

Now among metal packaging materials, the second priority comes to aluminum. The lightweight properties of the aluminum make this possibility. There are more desirable properties of aluminum, which enable it to be a good packing material of food products. They are optimum cost, resistive of corrosion, easy availability in nature, and most important is of recyclable property. Alumina does not have any need for protective coating material additionally to protect corrosion. This is because of the formation of the thin layer of aluminum oxide because this layer protects the parent materials for further corrosion when it comes into contact with oxygen.

Nevertheless, it is attacked by alkali metals. The disadvantage of using aluminum is that the cost is too high as compared to tinplate, but is lighter in weight. Aluminum is found in two categories in the packing materials. The first form is used as cans for packing of mainly beers and soft beverages. The second form is like aluminum foils. However, the purest form of aluminum is of ductile foil type, which is used for laminations of food products.

GLASS

A packaging material as glass containers is considered as the most renowned packaging means. It is mostly preferred for the packaging of wines, liqueurs, perfumes, and cosmetics items. The properties that make it enable to be favorable material for packaging. Those are its highly inert behavior, impermeable to vapors and gases also, and can behavior to any shape and size, i.e., easy to go for different shaping. There are so many benefits of glass as packing material. Those properties are mentioned below: transparent, but also can be given to any color as per the desire. Inert, impermeable nature, rigidness, thermal resistive, fits for the general consumer appeal, the most important thing is that it has selective light protection qualities. If it is focused on the compositions of the glass materials, then it is found that the bottle includes 68-73% of SiO_2. 12-15% of Na_2O, 10-13% CaO, and it has lesser proportions for the other oxides. Glass materials have better barrier properties for oxygen molecules, and it is entirely neutral when it comes to direct contact with the food particles. These days it can be a concluding remark about the glass materials is that it can be recycled again and again. However, it has also some disadvantages. They are brittleness, heavyweight, and it requires more energy to be manufactured. Glass can be recycled, but it has some difficulty in re-use. So, recycling is economically viable and technically also.

PAPER

For packaging purposes, paper products are used these days broadly. The use of paper bags for packaging materials is from the long term ago around the seventeenth century. Paper is defined in terms of its dimensional property, i.e., the sheet thickness is less than 0.23 mm, and following the weight, it is lighter than 220 g/m^2. The production of the paper and board materials are done from woods, rags, and other waste material. Firstly, it is treated so that it can be broken the lignin structure with the help of calcium bisulfite or using caustic soda. For a while, the paper material is decomposed first with the technique of bacterial action. As we can count this paper in the category of environment-friendly. But nowadays, due to the excessive use of plastic material for packaging application paper material is facing much competition in packing material in an open market.

The main advantages are low costing, readily anywhere available, lightweight, ability to print on itself, also having mechanical strength. However, due to some challenging problem of its property that the paper material is facing is turned into its disadvantage. Furthermore, those are low strength, resistive for water and gas (moisture). These disadvantages can be reduced up to more extent with some applications and treatments. Like, its nature towards moisture, i.e., its permeable nature for moisture and fat, can be minimized with a wax coating on its surface. The most markable thing is that now a day's paper is still used in the second priority for packaging, e.g., cardboard boxes or cartoon are the best examples of it.

PLASTICS

The most popular material used in these days in the food packaging industry is polymers. It was first in 1939 when the plastic material is entered into commercialized production. The reason behind its popularity for packaging purposes is its diverse nature and having broad-spectrum properties. Even more, plastics have unique qualities like, cheaper than other packing materials, can be easily shaped and processing is effortless, light in weight, very comfortable to seal, transparency or opaque. Plastics containers can be made of any shapes and sizes. The good thing is that their cost is too much lesser than the metals, glasses. If it is compared with the paper materials, then its density is higher than of it, but its density is lesser than half of the glass and aluminum density. The most relevant aspect of research work, which is going on currently is about the transport property in food packaging for the polymeric materials. Although metals and glasses do not show the behavior of permeability, polymers show their permeable nature for small particles. There are two limitations while using polymers as packing materials. They are permeability to gases and vapors, and another one is the migration of the packaged materials to the food products.

SHORTCOMINGS OF EXISTING PACKAGING MATERIALS: OPPORTUNITIES FOR NANOTECHNOLOGY

The concept of food packaging has been evolved its response towards the continuous development in the sector of materials science and technology. It is changing day per day according to the need of consumer's demand. In the present economy of the global world, the concept of packaging is not only to provide the effective distribution and food preservation assurance, but it is also concerned about communication and facilitation to their convenient end-use at all level of consumers who are using it.

The use of non-biodegradable based plastic materials for the food packaging application has brought a concerned issue of environmental problems of waste disposal. Day by day increasing demand for the quality improvement for the food products is the major key factor for the increasing rate of rapid development in the sector of biodegradable based nanomaterials for food packaging (Cutter, N.C. 2006)

The industries of food packaging are facing challenging problems in the sector of food technology. These problems are due to the lacking behavior of essential properties, and their absence made some opportunities for their innovations. If we try to enlist those needful properties that they are lacking can be listed as: (i) production materials are of non-sustainable type (ii) absence of the property like recycling (iii) not fulfilling the properties like such strength of mechanically and also barrier properties (Akbari, Z et al.,2007)

It is a well-known fact that about the plastics materials and metals that they have sufficient barrier properties to restrict the mass transfer of unwanted particles, which leads to the toxic behavior in food particles. However, their degradation is not possible in a biotic manner, and they are responsible for an unhealthy environment. Plastic materials are still accessible for the food packaging these days, also up to a percentage of 40% (Rhim, J.-W 2013). Because of their property of lightweight materials, formability, cost adequacy, and adaptable attributes. However, most of the food packaging materials come into the category of petroleum-based and which are non-sustainable if we consider the parameter of supply standpoint. The result is that the current scenario of North America is that out of the total amount of the municipal solid waste generated consists of 30% of the waste packaging materials, pressed the environmental issue. On the other side, if we talk about the barrier properties of water vapor and gases, then these packaging materials show weak behavior towards the barrier (Arvanitoyannis, I. S., & Bosnea, L. A., 2001). For example, packaging of live foods such as; green vegetables, fruits, and fresh products requires materials that permit O_2 through the package and behaves as permeable transmission materials having an optimal rate of transfer. However, processed products do not exhibit these types of mass transfer phenomena. Here, the most challenging thing is about barrier properties. The question arises here on how to ensure the availability of such thermoplastics, which have suitable matching barrier properties according to the specific kind of products? It is all about to increase the shelf life of the food products in contact with such thermoplastics. But to get rid of this problem, it has been developed the properties inside materials to expand their functional behavior like the thermoplastics, and for this polymer blends and multi-layered composite-like structure have been developed. The problem is remained unsolved due to the problem of costlier materials and also facing difficulty in recycling the material as mentioned above type properties. These days, manufacturers of the food products are facing trouble to set a balance between both the requirements of the society. It is about to get enough shelf-life for the product to be prepared with the best quality as well as a matter of the safety of the food products and consumers also. These problems are concerned about the countries which are now in the condition of the underdeveloped countries due to which they have no such technological development in regards to the distribution of foods and infrastructures for the preservation the food products.

There are some critical issues related to safety concerns and the quality of food products. Moreover, in the current scenario, there is an urgent need to overcome these lacking properties so that it should be clarified about their quality assurance of foods. They are listed below: (i) expansion of microorganism because of pollution and temperature abusement (ii) Due to the oxidation phenomenon, it has been noticed a decrement in the quality of nutritional amounts. (iii) Naturally, the interactions of the food particles with the solar light, oxygen content, and aqua particles used to change some of its intrinsic behavior and the resulting loss of organoleptic, i.e., qualities related to the nutrition system. So, the drawbacks, as mentioned earlier, are existing shortcomings that have to remove. So it can be assured 100% for the quality assurance of the food products as well as for human health. Over the previous decade, a lot of research work and development is being carried out for the use of nanocomposites in the sector of food packaging applications.

INNOVATIVE FOOD PACKAGING

During the storage and distribution of food products, food packaging technology provides us with protection and containment of the foods. This technology helps to keep away from the unwanted materials that may harm or can change the quality of the product. The unfavorable conditions may be in the category of the water vapor, released gases, orders, mechanical shocks, vibrations that may occur inside them, born microorganism, etc. (Duncan, T. V. 2011).

In modern society, due to complexity like human beings and their ongoing busy-life food manufacturers and industries are more concerned about the development process of the packaging concept system. They are continuously making an effort to add up more enhanced and convenience features according to the demand and also suitable for the safety issues of the food products, human beings, and environment. Orders, mechanical shocks, vibrations that may occur inside them, born microorganism, etc. (Lim, L. T. 2011).

There is great importance to innovations in the food packaging industry. These innovations in the packaging industries of food products have made a variety of new terms related to the aspects of safety, the shelf life of the food products, and the food product convenience. The surrounding climate of the sustenance inside the packaged product has a significant impact on the period of the product, i.e., shelf-life of the product. So, the principles needed to be followed while packaging products for maintaining the inside environment healthy for food products and better preservation are classified as; (i) passive, (ii) active, and (iii) intelligent packaging (also called sometimes smart packaging). The active system packaging itself consists of two types of packaging concepts; they are simple and advanced packing system.

Moreover, the intelligent packing system itself includes two other types of packing concept. They are termed as simple and interactive packing system.

Passive Packing System

It is defined as a packaging system of the passive type, which provides a barrier physically in between the food product and the surrounding environment around it inside the package. They are supposed to protect the food in a possessive way.

If we consider the conventional type of packaging system for food products, then most of them come into the category of the passive packing system. Examples of the passive packing systems are metal cans, glass, bottles, many more of the packing materials, which shows its flexible nature provides the same physical barrier in between the food products and the environment around it inside the packaging. This passive packaging system ensures the migrations that usually happen with the food products when coming in contact with the packing materials. So, the passive system provides prevention of coming in contact with the environment properties and the agents which are contained in the same environment. In a general way, the expectations from this packing system are to get maximum protection of the food products, but the fact is that it is not more responsive to the properties of the container.

In these days, the development work has been going on for the generation of newer coating materials for the polymeric containers and films also. Here, the question arises that why only these coatings are focused on implementing in container body? The reason behind this is that they have full control over the permeability of the migrative agents, which raises safety issues of the food products and also shows a negative impact on the shelf-life of the food products stored inside the container.

Active Packaging

An active packaging system can get information about whatever changes are going inside the environment of packaging products. It sends the feedback regarding this change through its sensing device. This packaging system is supposed to provide full protection as well as the preservation of the foods with the help of some intrinsic/extrinsic mechanism factors. It is defined as the packing, which helps to modify or extend the life span of that food product and gives more safety concerns about the food product as well as human health.

Simple Active Packing

A food packing system that does not include the presence of a functioning ingredient or effectively functional polymer is termed as a simple active packing system. An active system shows its response to any changes that are going within the food packaging due to the transfer mechanism of the materials to the food particle.

There is a kind of active packing system which is called modified atmosphere packing (MAP). It has been defined as a packing technique in such a way that it provides a modified atmosphere in comparison to the surrounding air composition. A very quiet natural definition of it is that it includes both kinds of packing behavior like; vacuum packing and (controlled atmosphere packing (CAP). So the most accurate definition comes for the MAP is "providing and continuously improved the environment in which the proportion of air is as per the desired nature of food within the package. Initially, the product is kept in the desired mixture of gases, and its composition depends on the type of food product, the packaging material, its average life-span, and the storage atmosphere. So, it can be added a remarkable point that whatever is the atmospheric condition within the package is mostly the outcomes of the result of the respiratory result as formed by the food products, selectively permeable, and the presence of added modifiers.

An example of a MAP system is about packing films that help to fulfill the desired atmospheric condition in maintaining the concentration of O_2 and CO_2 for the vegetables and fruits packaging. For categorize, the product type which falls under this category is the products of dairy, bakery, meats, and muttons, poultry items, fishes, and fresh green vegetables and fruits.

Advance Active Packing

An advanced active system is a type of packing system which contains polymers having actively functional behavior or having some active ingredients. In this packaging system, advance principles are being used in the packaging material. Those principles are listed below: (i) atmosphere modifiers (ii) absorber particles of O_2 molecules (iii) CO_2 generators and the absorber particles (iv) moisture regulators and ethylene absorber particles. The advanced active system of packaging has divided into two categories of the packaging system. The oxygen scavenging action for the food packaging is an excellent example of a system, which can absorb the unwanted foreign particles generated within the packaging environment when it comes to the contact of the food particle.

MAP system is another excellent example of the advance active packing system. It provides a barrier property to maintain that level of steady-state condition of the inside atmosphere proportion of air properties together with released gas due to the presence of food particles.

For the removal of oxygen, iron oxidation is used. Similarly, it is imperative to remove ethylene from the packaged material for the prevention of accelerated ripening effects. This ethylene is being removed by adding some active carbon content, or it can also be done by the oxidation phenomenon by $KMnO_4$. Similarly, cyclodextrin is the most emerging material for the scavenger of unwanted particles present during the storage of food items. The development of antioxidant film properties of protecting the shelf-life of the food product can be a significant factor in improving the attributes related to sensing issues inside the packaging products.

Intelligent Packaging

A packing system which exhibits the properties sends the feedback or information of the changing environmental condition and take corrective action to protect the food products during storage is called an intelligent packaging system. The identification of the intelligent packing system is to continuously enhance the aspects of the communication, i.e. feedback response of a package. So that it can be known about the situations inside the package about food quality in real-time. This intelligent packing gives an investigation report regarding the dating approach of "Best Before" and "Use by."

The feedback response of the intelligent packing system not only gives the information regarding the safety and product quality but also can be useful for the manufacturers to make decisions for the support system to verify questions like when and what steps should be taken in order to make the distribution of the entire product channel and production process.

So, there are some primary objectives of the intelligent packing system which has been enlisted below:

1. More improvement in the quality-based and product value of food materials.
2. Increased convenience manner
3. To bring the changes in the permeability properties of the gas
4. It should be assurance regarding the protection against additional effects like; tampering, counterfeiting, and theft.

The intelligent packing has provided some additional properties for an effective packing system. It includes oxygen indicators, can detect the pathogenic micro-organism and spoilage phenomenon, indicators giving information of the time,

Figure 2. Nanotechnology development and its implication in active and intelligent packaging

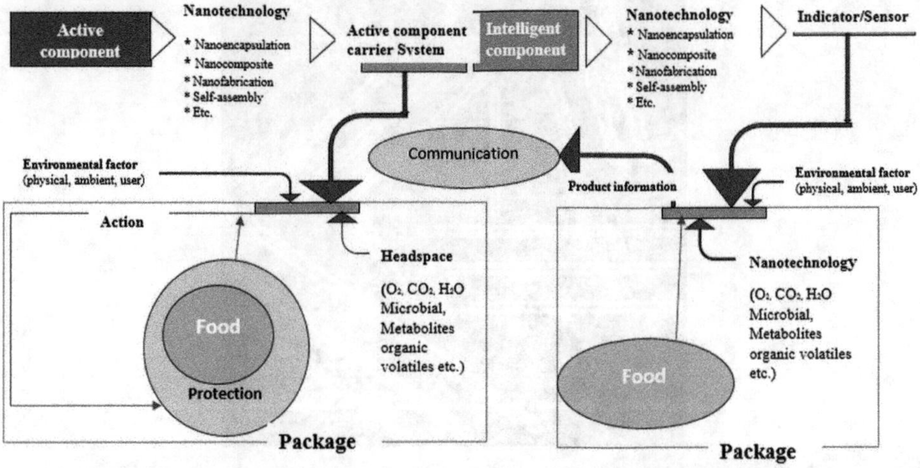

temperature and humidity inside the package. It also gives information about freshness providing real-time data of storage and distribution. A temperature indicating labels has been attached to the packaged external surface. It also gives information about the limit of the maximum temperature that has been exposed to the outer surface. It also has a component of the internal indicator having information about the current level of gas inside the packaging. These indicators are for monitoring the level of oxygen (O_2) and carbon dioxide (CO_2). To get the information regarding the contents of the food product, weight, locations, and timing throughout the channel distribution, it has been attached a Radio Frequency Identification (RFID). In the future, it may be possible to attach RFID levels for every food product. A schematic figure is shown in figure 2.

DEVELOPMENTS OF NANOCOMPOSITE FOR ENHANCED MECHANICAL AND BARRIER PROPERTIES FOR FOOD PACKAGING

Nanomaterials are known only due to their unique functional properties. It has a much larger surface to mass ratio as compared to larger-sized bulk materials. Nanocomposites are being incorporated with the polymer matrix to that substance due to their higher value of the surface area. Moreover, this increased surface area

Figure 3. Nanomaterials as building blocks for enhancing the mechanical strength

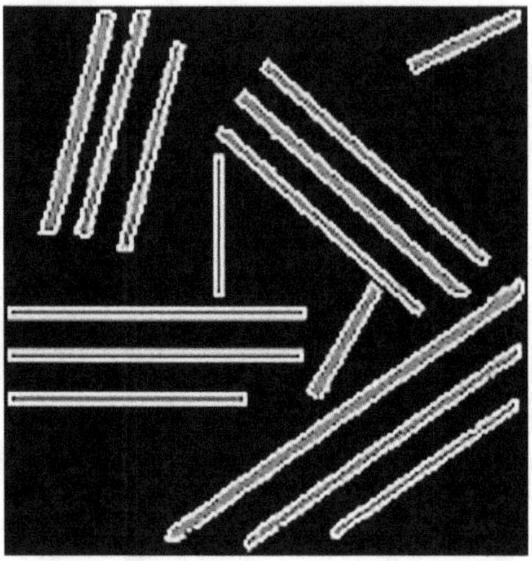

Figure 4. Nanomaterials providing plentiful surface area for filler matrix

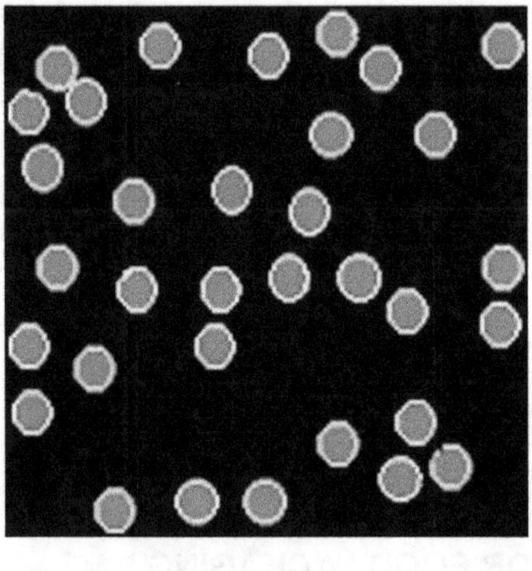

favors the filler-matrix, which is interacting in between and improving its performance. For the development of barrier properties introduction of nonreinforcement has been done, which provides a small barrier for the gases by complicating the path of that materials (Figure 3 & Figure 4).

Barrier Protection

The protection of food products can be done by maintaining an inert and having a low oxygen content atmosphere within the packaging bags. This inert behavior of the inside environment provides a barrier to the exchange of gases. This barrier results accurately and provide a healthy environment and maintain the quality of food with the help of clays. The antimicrobial packaging concept terminates the formation of the bacterial growth and fungal organism or any pathogens and toxins inside the package. These antimicrobials are formed with the help of silver oxide, titanium oxide, and zinc oxide, or it may be other bio-nano particles or nano-composite material.

NANO-BASED SENSORS

Sensors are the devices that can be used to detect any changes that have been occurred in the system boundary of food packaging. These changes might be possibly related to the physical, mechanical, chemical one. So. The sensors detect that changes occurred within the environment of the food packaging system and convert these changes into some readable form of observable signals. In this way, sensors regulate all its internal environment of the foodstuffs and its properties are continuously sensed and indicated by sensors. These changes belong to temperature, humidity, levels of oxygen content, degradation of food products and contamination of microbial organisms.

When the nano-based sensors are being used in an integrated form with the food packaging, then it is quite easily possible to detect specific compounds, pathogens available, toxins, to eliminate the inaccuracy in expiration dates, to provide the real-time data for fresh products of foods.

There are some of the examples of sensors that are being used for food packaging. A most popular type of metal oxide gas sensor is used due to its high sensitiveness and stable nature. Conducting polymer nanocomposites (CPC) or metal oxides are one of the useful sensors used in food packaging for the identification of gas emissions due to micro-organism. TiO_2, SnO_2, are some examples of O_2 indicating sensors. The color response plays here an important role in sensing the O_2 exposed. So, it is bleached when it has not been exposed and blue when being exposed.

The current packaging concept is fulfilled with the oxygen scavengers, moisture absorbers, smart barrier packing. Moreover, for bakery and meat products, the pacing concepts are enabled with nano-enabled sensing elements continuously giving updates regarding the inside environment of packing. Not even this, nowadays nanosensors also gives the information related to the shelf life of the food products.

EXTENSION OF SHELF-LIFE OF FOOD PRODUCTS USING SILVER NANOPARTICLES AND NANOCOMPOSITES AS ANTI-MICROBIAL FOOD PACKAGING MATERIALS

After long-term research work, it has been observed and concluded that silver particles could perform itself as an antimicrobial agent in the food industry and also have applications like beverage storage. In ancient times silver made vessels have been used for the storage of wines and water. Russian MIR space stations were using silver as a sterilization agent for water. For the developing countries, silver's property of behaving like broad-spectrum antimicrobial and also relatively lower cost than others made it possible as an active disinfecting agent treated for water.

Silver particles dominate other anti-microbial agents. It is only due to its additional properties, and they are (i) self-stable (ii) having a broad spectrum (iii) effective at penetrating biofilms (iv) effective bactericides. Silver behaves like an anti-microbial agent, which prevents the growth of microbes. It has been found that silver nanoparticles (AgNPs) serve as potent agents against bacteria species. It has also been observed that AgNPs are very effective against the strains of an organism, which shows its resistive nature towards the potent chemical antimicrobials. It is a well-known fact that inorganic nanoparticles have some more advantages than the molecular antimicrobials. Furthermore, this is possible only due to the controlled release of the AgNPs. So, it can be concluded that AgNPs/polymer nanocomposites can be applicable in both the medical unit devices as well as for food packaging materials to increase the shelf-life in a preservative manner.

AgNPs nanocomposite material offers more stability and has some relaxed release of silver ions into the food products. Another polysaccharide named chitosan, if it is loaded with the AgNPs, then it has some more tensile strength and better gas barrier properties developed. Similarly, AgNPs proportions are there in the Chinese jujube fruits, which are stored in storage bags. On keeping the fresh melon cutting stored in the pads of AgNPs contains cellulose, then it has been lowered the counts of microbial growth.

Application Polymeric Nanocomposites as Anti-Microbial Properties

Polymer nanocomposite materials help a lot in controlling the growth of microorganisms inside the food packaged material. It becomes possible due to the desirable property of structural integrity and barrier properties, which is imparted by the polymeric matrix. It also enhances the antimicrobial properties due to the agents, which have already been embedded within it. Nanocomposites have a higher ratio of surface to volume, that is why they can attach themselves with more number of biological molecules. It leads to a higher value of efficiency when it is compared with its microscale counterparts. These materials are versatile so that it can be used in numerous means as antibiotic carriers, killing agents, or inhibitors, which helps in growing. (Shankar et al., 2016). Various research work has been enhanced its properties like; it has been noticed an increased efficacy against Escherichia coli when nanocomposite is with less silver as compared to micro composite with more silver content. (Shyam S. Sablani 2015)

Moreover, it is being concluded that when polyamide 6 is filled with a weight percentage of 2 of Ag-NPs shows a more useful nature against E. coli, even after immersing it into the water for 100 days. It has also been reported that it starts retarding the senescence of jujube, (a Chinese fruit) when PE nanocomposite film is mixed with Ag-NPs.(Emamifar A, 2011).

BIO-NANOCOMPOSITES FOR FOOD PACKAGING APPLICATIONS

Most of the materials used for food packaging comes in the category of non-biodegradable materials category. They are not fulfilling the increased demand of the society and also not so cared about the issues related to the sustainable development and safety of the environment. In order to reduce the dependency of the food packaging materials on fossil fuels, there are more innovative technologies are currently running so that it can be moved to a sustainable type of material. To enhance the quality of food and increase the shelf-life of the food products, a lot of efforts are going on to reduce the waste generations from the food packaging due to its non-biodegradable nature (Avella, M et al., 2005). The disadvantage of using non-biodegradable materials or plastics for packaging purposes is that these products have been derived from petroleum products. It creates a problem of waste disposal. Due to the weak mechanical strength and having not excellent barrier properties the application of biopolymers has been limited. Moreover, the scarcity of the desired property can be enhanced by the addition of reinforced nano-sized particles (Tang et al., 2012).

The establishment of the nanocomposite material has been done to improve the mechanical as well as barrier properties of the biopolymers for the food packaging. Bio-composites are multiphase-materials which have the continuous phase of the two or more constituent particles and having a discontinuity in nanofillers (<100 nm). The role of the nanofillers is more essential to improve the properties like mechanical and barrier. There are some examples of nanofillers (silicates, TiO_2, and clay) that not only improve the property mentioned above but; it offers some more critical functions. They are listed as an antimicrobial agent, bio-based sensor, and lastly, oxygen scavenger. At the earlier days, the application of biopolymers for packaging of food products were starch, cellulose, chitosan, and agar. The origin of the agar was from carbohydrates and gelatin, gluten, alginate, and collagen have been derived from protein. Recently, after a lot of research work, it became possible for bio-degradable polymeric nanocomposites to develop those properties which are more suitable for full applications in the packaging industries.

It can be concluded that for the packaging application, the most suitable biodegradable nanocomposite materials are polylactic acid (PLA), polybutylene succinate (PBS), polyhydroxy butyrate (PHB), aliphatic polyester (PCL) (Rhim, J.-W. et al., 2013).

Application of Bio Nanocomposites for Food Packaging

The extraction of the biodegradable packaging material is done from the resources of bio-based renewables. These materials are generally called as biopolymers having much good barrier and mechanical properties. They degrade biologically at the end of their life. These polymers have been evolved as an environmentally friendly substitute option for the non-biodegradable plastic materials for packaging purposes. (Pan et al., 2014).

Among most of the bio-polymers used for food packaging purposes are carbohydrates and their derivatives. Research shows that poor mechanical strength of starch can improved with mixing additives like; plasticizers, nanofillers. (Sorrentino et al., 2007).

Other biopolymer-based research work has been performed with cellulose acetate (Bruna et al., 2014), alginate (Shankar et al., 2016), pectin (Shankar et al., 2016c) synthetic biopolymer as PLA (Conte et al., 2013;), PHA (Bordes et al., 2009), PCL (Gorrasi et al., 2013, and microbial synthesized polymers such as; PBAT (Shankar and Rhim, 2016b)., agar (Orsuwan et al., 2016), chitosan. (Pereda et al., 2014).

PLA Nanocomposites

PLA is the most common bio-based nanocomposite material known for its biodegradable nature in the environment. The production of the PLA is done from the lactic acid. It is a simple process that is produced through the fermentation process of the carbohydrates originated from plant resources. Some of the examples are sugar beet, molasses, wheat and corn (Sorrentino, A et al., 2007). The advantages are self-biodegradability and can be set as a personalized life span as per the requirements. There are mainly two forms of the lactic acid monomer (L form or D form). The properties of the L type PLA material are that it has a higher value of the melting point and also have higher crystalline behavior, whereas another form is a mixture of D and L- PLA has a lower value of transition temperature comes into the category of an amorphous polymer. So far, the production of the PLA type bio-degradable polyesters for the packaging material have some limitations regarding higher cost, and its performance is not so good.

PCL Nanocomposites

Polycaprolactone (PCL) is manufactured by the process of ring-opening polymerization of ε-caprolactone. This type of polyester has a linear shape and comes in the category of a semi-crystalline polymer. The level of crystallinity is around 50%. At the breaking point, it shows its mechanical behavior like high elongations at breaking point and low modulus value. Nowadays, PCL is getting popular only due to its excellent physical properties and its availability in a commercial manner for commodity application.

Starch Nanocomposites

After excellent research work, it has cleared that for the packaging of food products, Starch is the right choice for its utilization. The properties that enable starch for this capability is possible only due to its eco-considerate compliancy, a more extensive range of availability, cheaper rate. The resistance for the water has enhanced with the addition of matrices of the synthetic polymer and the use of inorganic materials.

Recently investigation has been performed on the blends of starch with the nanoclays. It has been reported from this investigation that it is compatible with food packaging due to its biodegradable nature and versatile applications. It has also been noticed an enhancement regarding its mechanical behavior. The most important thing is that consideration of MMT is responsible for enhanced tensile strength and modulus of elasticity. It has also minimized the chances of absorption of water molecules, and it becomes possible with the presence of MMT in starch.

Recent data shows that as the concentration of the ZnO-CMC is increased from 0 to 5% by weight, then it leads an increment of 3.9 to 9.8 MPa in tensile strength, while there is a decrement in the elongation from 42.2% to 25.8%.

Cellulose Nanocomposites

Cellulose is an exceptionally strong natural polymer that exists in nature, and these are the building blocks of the long fibrous materials. The production of cellulose is done sustainably from the biomass and the places from the availability of natural polymers. The properties of the cellulose nanocomposites are so attractive that it makes them accessible in the application of food packaging. They are as follows: Lower cost, very high mechanical strength, the specific surface area is bigger, higher aspect ratio, lower value of density, lower thermal coefficient of expansion, tensile strength is good (Metzger, C. et al., 2018)

Cellulose nanofibrils (CNFs) are efficient nanomaterials because it acts as a barrier protection against the residue of mineral oil and oxygen content also. In a similar way methylcellulose (MC) is derived from the nanocomposites bio-film can minimize the challenging problem related to the environment.

Protein Nanocomposites

In recent years, the research has been focused on enhancing the properties of the protein on the film edible and coatings for the packaging material. These are the proteins that have been obtained from animals and can be utilized commercially. Most of the utilized zones are caseins, collagens, egg white, whey protein, and myofibrillar proteins (Jimenez. A et al., 2018), (Rocha. M et al.,2018)

Protein nanocomposites do not have good resistance ability for moisture content, but it has higher rigidity. Moreover, nowadays, whey protein isolate (WPI) has opted as an edible film as the best option for the food packaging materials (Sanyang M et al., 2018).

Recently, it has been derived from a hydrophobic protein from the corn kernels and named Zein. It has inherent characteristics of film formations. In the sector of food industries, Zein is used as coating substrate materials that have an inbuilt property of the biodegradable polymer. Soy protein is also a type of nanocomposite material when it is filled with nano-cellulose which is originated from a residue of licorice. This gives assurance for the sensitivity towards the water and also to increase mechanical strength (Orsuwan et al., 2018)

SAFETY ASSESSMENT OF NANOMATERIALS IN FOOD PACKAGING

The primary concern about the use of nanocomposite materials is how to ensure customers and environment also about its only global perspectives suitable for food packaging. To ensure its suitability and perfectness, it needs excellent evaluation for its properties and positive response. So, in that direction, research work needs to fulfill five markable questions to ensure that nanocomposite and nanotechnology is the better material technique for food packaging. Those cross-questions are enlisted below:

1. Firstly, how to determine or how to examine the presence of nanomaterials in the sector of food packaging sends potential hazards when they come into direct contact with the oral route? Because it is applicable for all nanoparticles used, so it is not a lack of knowledge issue, but any negative impact on the quality of food will raise the questions of hazards material. So, research needs to think over it seriously to overcome it (Simon. et al., 2008).
2. The second question is concerned about lacking the tools that tell us the exact exposure. That is, if the migration phenomenon is going on, then how much proportion is limited about it? Is it utterly depending upon the concentration of the nanoparticles or on the number of particles present? Is it related to the size and shape of nanoparticles? So, how hazard type is being characterized? (Teide et al., 2008)
3. The third one is related to surface chemistry. It asks that is it the only reason for unwanted chemical reactions, which is due to the presence of nanomaterials. It is surface chemistry that gives elevation to the migration phenomenon which results in the formation of unwanted products during the fabrication and processing of the packaging materials (Teide et al.,2009).
4. The fourth question is all about what negative impact does nanomaterials shows towards the waste disposal streams. The concept of recycling and reuse, burning for recovery of energy, and lastly, a viable and straightforward arrangement for landfilling. Which option will be better whether the waste disposal system or recycling process? The matter is that when they are allowed for the transformation process, then they will able to retain their behavior like physical, chemical and toxic behavior or not? (Simon P et al.,2008)
5. Last but not least, the fifth point is about the real issues and worries on nanomaterials that they have cast a shadow on the conventional technologies for food packaging. A direct question arises about the differences between the already determined and has permission conventional technology and the

novel one. In this manner, where and how the overlapping is happening in conventional technology?

Research only needs the answer to these above mentioned five critical questions to do the hazard assessment and its characterization successfully.

HAZARD ASSESSMENT OF ENGINEERED NANOPARTICLES IN FOOD: NANO-RISK CAT TOOL APPROACH

For the assessment of hazards and exposure potentials due to the presence of the nanomaterials in the food products, a systematic approach has been developed, which is named as a nano-risk Cat tool approach. It gives its analytical results in the form of dot indications. Moreover, all dots have different features related to potential hazards. The three consecutive dots representing for the professional users, for the consumers, and the third one is for the environment. The other two dots used for the denotation of the potential hazards for the human being and environment respectively. This technique can be used for understanding and distinguishing the exposure of known potential hazards, which will be caused due to the application of nanoparticles for food packaging. The essential features of this technique are that in case of where there is a lack of data that hampering the traditional risk assessment procedure. This analysis is only possible when there is the presence of the expert judgments opinion which can give their response based on the dots coming (EFSA., 2011)

The migration phenomenon of the nanocomposites packaging material in contact with the foods is one of the most critical worries for humans. Therefore, issues of health risks come. Furthermore, the effect of toxicity behavior present in the nanomaterials mainly depends upon the fact of how much concentration of food is consumed (Chen et al., 2013).

The researcher's considerations have been pulled towards the safety-related issues for nanocomposites due to their enormous use as a packaging material in the food industry. In the present age of commercialization, worldwide organizations are continuously giving their effort to gather information about the current health and safety risk associated with the use of nanocomposite materials for food packaging and production. These organizations are also concerned about the migration of materials when it comes in contact with the foods.

It can be easily seen from our daily life examples, that how these hazards quite naturally come, that we did not bother about it. Therefore, here are some enlisted examples who are responsible for all these things to happen, which are related to the health issues of human beings. Likewise, a person chewing the gum admitted into his body about 93% of TiO_2 of the nano-sized particle. Because it contains the same proportions into the chewing gum sugar-coated. Similarly, consuming foods that contain E551 allows inhalation of gut epithelium-like materials, which has very much tendency to be exposed towards SiO_2 nanomaterials (Margo et al.2014).

MIGRATION ASSESSMENT

Allergies

Two considerable critical issues show its adverse effects on exposure with nanomaterials are an allergy, and another one is massive metals release. As the food technology has been reached up to the right level of research work, still it cannot be ignored with the fact that the presence of nanoparticles can promote hypersensitive pneumonic inflammation in coming in contact with the food items. Moreover, from the previous research work, it can be revealed that due to their presence, the inflammatory response, and an increment in the ROS production, it has become a typical immune response with the exposure of nanomaterials. Even carbon contained nanomaterials are also responsible for the cause of allergic inflammable. It has also been found in the research work that carbon nanotubes of a single and multi-walled type tend to increase the lung type inflammation. A specific type of allergen level of IgE type of mice sensitized to the OVA-specific egg allergen.

Heavy Metal Release

It is a proven fact that the presence of metal-based nanomaterials not only encourage the mechanical strength, but it also provides barrier properties, and prevents the phenomenon of photo-degradation of plastics, and shows its properties like an effective antimicrobial in the form of bulkier ions of metals. However, on the other side, it has a contradictional argument about its toxic behavior. The release of heavy metals causes toxic outcomes due to the presence of nanomaterials. It is only due to long term contamination of the release of heavy metals into the food particle which shows an adverse effect. There are much more nanomaterials which are in the category of metal-based nanomaterials, but among them, three are of those type which is mostly recorded metal-leaching. They are ZnO, Ag, and CuO and their presence increase the level of intracellular ROS, which results in DNA damage and phenomenon like lipid peroxidation.

NEED FOR REGULATIONS

The safety issues of the nanocomposite materials in the food packaging are increasing rapidly; hence, attention has been attracted to the safety-related issues for both the human being as well as the environment. For the safety assessment, there needs to a granted market permission for the components like plastic materials for packaging purposes. As per the chemical and toxicity data submission through the manufacturer side, it is based on the guidelines of data requirement (Bradley., et al. 2011)

At present various world-wide agencies are working on getting the information regarding the health and risk related to the use of nanotechnology in the packaging of foods. An international organization WHO (World Health Organization) with FAO (Food and Agriculture Organization) (FAO, 2009) has given a recent report on the use of nanotechnology in the sector of food and agricultural growth in a meeting of a joint expert meeting which was held in 2009. It was totally about the potential food safety implications. Another establishment has been done by the organizational council for Economic Cooperation and Development (OECD) (OECD., 2009). The issues were related to the health of human beings and the safety aspects of the environment for nanomaterials in chemical sectors of food technology. Therefore, in the future, worldwide agencies are so much concerned about it. Continuously, they are raising the issues related to food safety, and their organizations are still working for the development of nanomaterials that are not harmful like the toxicity effect of nanomaterials.

The EFSA has been authorized to organize and give recommendations for the issues which govern the quality of the foods for the human being as well as the environment also. This regulation offers such recommendations only for European Commissions. These regulations provide us with information about the active components which are present in the nanomaterials used for the food packaging. It also gives realistic data about the maximized quantity that has been released from the active materials present in the nanoparticles (Schmidt et al., 2011). There is an urgent need for the overall safety of the environment from the disposal of packaging materials. Two paramount concern is there about global environment protection and second is saving the natural resources, need to develop environmentally-friendly packing concept. The concept is reusable, recyclable, and environmentally friendly disposal.

ASPECTS OF SAFETY TO HUMANS AND THE ENVIRONMENT

There are only two critical issues of significance related to the safety concerns of the use of nanocomposites materials for food packaging. The first and most preference goes to the safety and quality-based products of food and their impact on the consumer's health. The aspects of safety concern arise. The question is that the use of nanocomposites and technologies for food packaging and migration phenomenon puts negative questions impact the quality and assurance of the consumer's health.

Furthermore, if we talk about the second question related to the environment, if looked deeply into the environmental issues, then the first thing is about the manufacturing of the materials needed for packaging, but more important than this is how these materials are being disposed of when they have used up finished. This leads to the recycling concept of the nanomaterials. And another aspect is of compromising nature and performance quality of the product of nanomaterials after being recycled. What regulations and market uptake are obstructed by these uncertain natures belonging to the consumer as well as environmental safety issues.

The councils who defines some standard regulations keeping because of safety concern to protect human health and to prevent the adulteration of foodstuffs has set some limits of migration. It has been defined as two types of migration limits in case of the plastic materials used for packaging purposes for food (The Council of the European Communities.,1990). They are as follows:

1. The first one is about the Overall migration limit (OML) of the granted permission plastic materials when it comes in contact with the food materials is 60mg/kg of substances in any of the conditions.
2. Secondly, it is about the Specific migration limit (SML), which governs the migration level of materials that send hazardous toxic effects coming in contact with the food particles. These levels of migration are decided on the daily basis of tolerable daily intake (TDI), and another one is acceptable daily intake (ADI), which is highly recommended by the regulation of EFSA. These regulations also provide the procedures of testing like overall migration (The Council of the European Communities, 2007).

OPPORTUNITIES FOR DEVELOPING COUNTRIES

The advantages offered by the application of nanomaterials and nanotechnology are frequently realized in near-market applications. They include the following benefits: (i) innovations (ii) Light -weighing (iii) massive extent of protection and preservation of foods (iv) performance enhancement in the bio-based materials.

Nowadays, nanotechnology along with information communication technology (ICT), has emerged as a new technology for the developing country's markets. They are getting renovated or adapted with the new technology for storage and transport of agricultural products, animal products, and food products. For the developing countries, this real-life relevance is so much needful to the poor peoples.

CONCLUSION

The use of nanocomposites materials and nanotechnology offers new opportunities for the foods and agriculture-based industries. Nano-composite materials have greater surface area per unit mass and are more biological activity than the larger sized particles. Bio-nanocomposite materials have more improved mechanical, thermal, and barrier properties for gases within the food package. In place of conventional packaging materials, it is far better to use bio-degradable packaging materials. For packaging purposes, silver nanocomposites, protein, starch, cellulose, PLA, PCL based nanocomposites are being used these days as they are bio-degradable. So, they are environment-friendly and also satisfy the need for customer safety.

Nanotechnology has more significant potential to provide a better quality of food products by making the food items tastier, healthy, and safety for the consumers. This technology innovates some new food products and improves the way of packaging and storage. In addition to food packaging, nano-sensors holds a vital place for maintaining food preservation. They detect whatever changes are going on inside the packaged food product and regulate the internal environment of foods. They are responsible for the enhanced shelf-life of the food products. Nanotechnology provides the benefits not only within the food packaging but also around the packaging. The application of nanocomposites/ nanotechnology detects the presence of bacteria, shows healthy behavior and also shows color response as per the inside quality of foods. Nanomaterials provide secure barrier property for the safety of food products. So, it will be better to communicate the benefits of using nanocomposite materials in food packaging to increase the level of acceptance of consumers towards this technology.

FUTURE SCOPE & RESEARCH DIRECTIONS

For ideal packaging, most of the target is to satisfy the standards as well as the customers, which have been discussed in this chapter. However, still, there is a compromise between the choice of package and the desired objectives. In the development of food packaging, quality assurance, safety, freshness, and convenience are considered as future needs and targets. Future trends arise questions on the characteristics like; opening case, small portions, consumer safety, environmentally friendly packaging, methods compatible with tamper-proof, and reclosable packaging.

Oxygen scavenging, revolutionary advancements in all cases optimum condition should be fulfilled. In the future, the driving forces of more research work, more innovations will be generated from the following mentioned questions which are: Who and Whom should pay for the collections and sorting? Is packaging has been done excessively? What management should be done to reduce the condition of excessive packaging? So, for this, the consumer should be educated to understand all the needs/ aspects of edible food packaging. How to make the right partner in food packaging for future development? These concepts and thoughts can help to get safe packaging for the customers.

ACKNOWLEDGMENT

The authors sincerely acknowledge the suggestions and comments of the authors are heavily indebted to them in providing their valuable suggestions and insights.

REFERENCES

Akbari, Z., Ghomashchi, T., & Moghadam, S. (2007). Improvement in food packaging industry with biobased nanocomposites. *International Journal of Food Engineering*, *3*(4). doi:10.2202/1556-3758.1120

Arvanitoyannis, I. S., & Bosnea, L. A. (2001). Recycling of polymeric materials used for food packaging: Current status and perspectives. *Food Reviews International*, *17*(3), 291–346. doi:10.1081/FRI-100104703

Avella, M., De Vlieger, J. J., Errico, M. E., Fischer, S., Vacca, P., & Volpe, M. G. (2005). Biodegradable starch/clay nanocomposite films for food packaging applications. *Food Chemistry*, *93*(3), 467–474. doi:10.1016/j.foodchem.2004.10.024

Bordes, P., Pollet, E., & Avérous, L. (2009, February 1). Nano-biocomposites: Biodegradable polyester/nanoclay systems. *Progress in Polymer Science*, *34*(2), 125–155. doi:10.1016/j.progpolymsci.2008.10.002

Bradley, E. L., Castle, L., & Chaudhry, Q. (2011). Applications of nanomaterials in food packaging with a consideration of opportunities for developing countries. *Trends in Food Science & Technology*, *22*(11), 604–610. doi:10.1016/j.tifs.2011.01.002

Bruna, J. E., Galotto, M. J., Guarda, A., & Rodríguez, F. (2014, February 15). A novel polymer-based on MtCu2+/cellulose acetate with antimicrobial activity. *Carbohydrate Polymers*, *102*, 317–323. doi:10.1016/j.carbpol.2013.11.038 PMID:24507287

Chen, X. X., Cheng, B., Yang, Y. X., Cao, A., Liu, J. H., Du, L. J., ... Wang, H. (2013). Characterization and preliminary toxicity assay of nano-titanium dioxide additive in sugar-coated chewing gum. *Small*, *9*(9–10), 1765–1774. doi:10.1002mll.201201506 PMID:23065899

Coles, R., McDowell, D., & Kirwan, M. J. (2003). *Food packaging technology*. CRC Press.

Conte, A., Longano, D., Costa, C., Ditaranto, N., Ancona, A., Cioffi, N., ... Del Nobile, M. A. (2013, July 1). A novel preservation technique applied to fiordilatte cheese. *Innovative Food Science & Emerging Technologies*, *19*, 158–165. doi:10.1016/j.ifset.2013.04.010

Cutter, C. N. (2006). Opportunities for bio-based packaging technologies to improve the quality and safety of fresh and further processed muscle foods. *Meat Science*, *74*(1), 131–142. doi:10.1016/j.meatsci.2006.04.023 PMID:22062722

Driscoll, R. H., & Paterson, J. L. (1999). Packaging and food preservation. In Food science and technology, (pp. 687-734). Marcel Dekker.

Duncan, T. V. (2011). Applications of nanotechnology in food packaging and food safety: Barrier materials, antimicrobials and sensors. *Journal of Colloid and Interface Science*, *363*(1), 1–24. doi:10.1016/j.jcis.2011.07.017 PMID:21824625

EFSA Scientific Committee. (2011). Guidance on the risk assessment of the application of nanoscience and nanotechnologies in the food and feed chain EFSA Scientific Committee. *EFSA Journal*. doi:10.2903/j.efsa.2011.2140

Emamifar A. (2011). *Applications of antimicrobial polymer nanocomposites in food packaging*. Academic Press.

FAO. (2009). *WHO Expert Meeting on the Application of Nanotechnologies in the Food and Agriculture Sectors: Potential Food Safety Implications.* Meeting Report 1-102OECD. Guidance Manual for the Testing of Manufactured Nanomaterials, ENV/JM/MONO 20/REV.

Gorrasi, G., & Pantani, R. (2013, May 1). Effect of PLA grades and morphologies on hydrolytic degradation at composting temperature: Assessment of structural modification and kinetic parameters. *Polymer Degradation & Stability*, *98*(5), 1006–1014. doi:10.1016/j.polymdegradstab.2013.02.005

Jiménez, A., Requena, R., Vargas, M., Atarés, L., & Chiralt, A. (2018). Food Hydrocolloids as Matrices for Edible Packaging Applications. In Role of Materials Science in Food Bioengineering (pp. 263–299). doi:10.1016/B978-0-12-811448-3.00008-5

Lim, L. T. (2011). Active and Intelligent Packaging Materials. In Comprehensive Biotechnology, Second Edition (Vol. 4, pp. 629–644). doi:10.1016/B978-0-08-088504-9.00308-1

Magro, M., Campos, R., Baratella, D., Lima, G., Holà, K., Divoky, C., ... Vianello, F. (2014). A magnetically drivable nanovehicle for curcumin with antioxidant capacity and MRI relaxation properties. *Chemistry (Weinheim an der Bergstrasse, Germany)*, *20*(37), 11913–11920. doi:10.1002/chem.201402820 PMID:25079005

Metzger, C., Sanahuja, S., Behrends, L., Sängerlaub, S., Lindner, M., & Briesen, H. (2018). *Efficiently Extracted Cellulose Nanocrystals and Starch Nanoparticles and Techno-Functional Properties of Films Made Thereof.* Coatings. doi:10.3390/coatings8040142

Mihindukulasuriya, S. D. F., & Lim, L. T. (2014). Nanotechnology development in food packaging: A review. *Trends in Food Science & Technology*, *40*(2), 149–167. doi:10.1016/j.tifs.2014.09.009

Miltz, J. (1992). Food packaging. In D. R. Heldman & D. B. Lund (Eds.), *Handbook of Food Engineering* (pp. 667–740). New York: Marcel Dekker.

OECD. (2010). *Guidance manual for the testing of manufactured nanomaterials: OECD's sponsorship program.* OECD.

Orsuwan, A., Shankar, S., Wang, L. F., Sothornvit, R., & Rhim, J. W. (2016, October 1). Preparation of antimicrobial agar/banana powder blend films reinforced with silver nanoparticles. *Food Hydrocolloids*, *60*, 476–485. doi:10.1016/j.foodhyd.2016.04.017

Orsuwan, A., & Sothornvit, R. (2018). Active Banana Flour Nanocomposite Films Incorporated with Garlic Essential Oil as Multifunctional Packaging Material for Food Application. *Food and Bioprocess Technology*, *11*(6), 1199–1210. doi:10.100711947-018-2089-2

Pan, H., Xu, D., Liu, Q., Ren, H. Q., & Zhou, M. (2014). Preparation and characterization of corn starch- nanodiamond composite films. In *Applied Mechanics and Materials* (Vol. 469, pp. 156–161). Trans Tech Publications; doi:10.4028/www.scientific.net/AMM.469.156

Pereda, M., Dufresne, A., Aranguren, M. I., & Marcovich, N. E. (2014, January 30). Polyelectrolyte films based on chitosan/olive oil and reinforced with cellulose nanocrystals. *Carbohydrate Polymers, 101*, 1018–1026. doi:10.1016/j.carbpol.2013.10.046 PMID:24299870

Rahman, M. S. (1999). Purpose of food preservation and processing. In M. S. Rahman (Ed.), *Handbook of food preservation* (pp. 1–10). New York: Marcel Dekker.

Rhim, J. W., Park, H. M., & Ha, C. S. (2013). Bio-Nanocomposites for Food Packaging Applications. *Progress in Polymer Science, 38*(10-11), 1629–1652. doi:10.1016/j.progpolymsci.2013.05.008

Rocha, M., Alemán, A., Romani, V. P., López-Caballero, M. E., Gómez-Guillén, M. C., Montero, P., & Prentice, C. (2018). Effects of agar films incorporated with fish protein hydrolysate or clove essential oil on flounder (Paralichthys orbignyanus) fillets shelf-life. *Food Hydrocolloids*. doi:10.1016/j.foodhyd.2018.03.017

Sanyang, M. L., Ilyas, R. A., Sapuan, S. M., & Jumaidin, R. (2017). Sugar palm starch-based composites for packaging applications. In Bionanocomposites for Packaging Applications (pp. 125–147). doi:10.1007/978-3-319-67319-6_7

Schmidt, B., Katiyar, V., Plackett, D., Larsen, E. H., Gerds, N., Koch, C. B., & Petersen, J. H. (2011). Migration of nanosized layered double hydroxide platelets from polylactide nanocomposite films. *Food Additives & Contaminants. Part A, Chemistry, Analysis, Control, Exposure & Risk Assessment, 28*(7), 956–966. doi:10.1080/19440049.2011.572927 PMID:21614708

Shankar, S., & Rhim, J. W. (2016, October 1). Tocopherol-mediated synthesis of silver nanoparticles and preparation of antimicrobial PBAT/silver nanoparticles composite films. *Lebensmittel-Wissenschaft + Technologie, 72*, 149–156. doi:10.1016/j.lwt.2016.04.054

Shankar, S., & Rhim, J.W. (2016). Polymer nanocomposites for food packaging applications. *Functional and Physical Properties of Polymer Nanocomposites*. Doi:10.1002/9781118542316.ch3

Shankar, S., Tanomrod, N., Rawdkuen, S., & Rhim, J. W. (2016, November 1). Preparation of pectin/silver nanoparticles composite films with UV-light barrier and properties. *International Journal of Biological Macromolecules, 92*, 842–849. doi:10.1016/j.ijbiomac.2016.07.107 PMID:27492557

Simon, P., Chaundry, Q., & Bakos, D. (2008). Migration of engineered nanoparticles from polymer packaging to food - A physicochemical view. *Journal of Food and Nutrition Research, 47*(3), 105–113.

Sorrentino, A., Gorrasi, G., & Vittoria, V. (2007). Potential perspectives of bio-nanocomposites for food packaging applications. *Trends in Food Science & Technology, 18*(2), 84–95. doi:10.1016/j.tifs.2006.09.004

Tang, X. Z., Kumar, P., Alavi, S., & Sandeep, K. P. (2012). Recent Advances in Biopolymers and Biopolymer-Based Nanocomposites for Food Packaging Materials. *Critical Reviews in Food Science and Nutrition, 52*(5), 426–442. doi:10.1080/10408398.2010.500508 PMID:22369261

The Council of the European Communities. (1990). Relating to plastics materials and articles intended to come into contact with foodstuffs. *Official Journal of the European Union, L, 349*, 26–47.

The Council of the European Communities. (2007). *Relating to plastic materials and articles intended to come into contact with food and Council directive,* 85/572/ EEC laying down the list of simulants to be used for testing migration of constituents of plastic materials and articles intended to come into contact with foodstuffs. *Official Journal of the European Union, L, 91*, 17–36.

Tiede, K., Boxall, A. B. A., Tear, S. P., Lewis, J., David, H., & Hassellöv, M. (2008). Detection and characterization of engineered nanoparticles in food and the environment. *Food Additives & Contaminants. Part A, Chemistry, Analysis, Control, Exposure & Risk Assessment, 25*(7), 795–821. doi:10.1080/02652030802007553 PMID:18569000

Tiede, K., Hassellöv, M., Breitbarth, E., Chaudhry, Q., & Boxall, A. B. A. (2009). Considerations for environmental fate and ecotoxicity testing to support environmental risk assessments for engineered nanoparticles. *Journal of Chromatography. A, 1216*(3), 503–509. doi:10.1016/j.chroma.2008.09.008 PMID:18805541

Chapter 6
Recent Advancements in Photocatalytic Nanocomposites

Aruni Shajkumar
https://orcid.org/0000-0001-6958-5236
School of Advanced Research in Polymers, Central Institute of Plastics Engineering and Technology, India

Ananthakumar Ramadoss
https://orcid.org/0000-0002-6616-7693
School of Advanced Research in Polymers, Central Institute of Plastics Engineering and Technology, India

ABSTRACT

Photocatalysts utilize light energy or photons to catalyze a reaction. The most significant characteristics of a photocatalyst lies in its ability to simultaneously oxidize a donor molecule and reduce an acceptor through the electron-hole pair generated upon excitation. With the emergence of nanotechnology, the utilization of nanomaterials for their photocatalytic properties has gained a new pace. TiO_2 *and ZnO nanoparticles are exploited widely for their photocatalytic properties. The recent trends concentrate on devising composite nanostructures that utilize both the properties of the photocatalyst and supporting materials such as graphene, carbon nanotubes, or noble metal nanoparticles to enhance the photocatalytic properties of the semiconductor metal oxide. The main areas of application of such structures lie in the field of water purification and energy harvesting. This chapter outlines an overview of the photocatalytic process and the existing technologies followed by the application areas and the recent advancements lying in that area.*

DOI: 10.4018/978-1-7998-1530-3.ch006

INTRODUCTION

Environmental pollution is one of the major challenges that the earth is facing nowadays. Increasing population and overuse of pollutants to 'facilitate an easy life', adversely affect the natural resources such as water, air and soil. Among the many pollutants, organic pollutants play a major role in causing water pollution. Majority of these pollutants come from industrial effluents, plastics, detergents, organic dyes, synthetic fibres and rubber, paints, food additives, pharmaceuticals and other organic wastes (Bhanvase, 2017). These pollutants get introduced not only into surface water but also to groundwater. Therefore, researchers are in a quest for an effective method to treat such pollutants. Among the different available methods, photocatalysis is gaining a lot of interest due to the capability of photocatalytic materials to harvest solar energy and undergo reaction at mild reaction conditions. Photocatalysis is the process that involves the utilization of a light-sensitive material to catalyze a reaction. It can be classified as an up-hill process or down-hill process, where up-hill reactions convert light to chemical energy whereas down-hill reactions convert reactants to products using light energy. Semiconductor metal oxides are widely used in photocatalysis due to their innate photosensitive properties (Ashkarran, 2015; Bhanvase, 2017; Divya, 2017; Hatamie, 2015; Li, 2007; Liu, 2018; Lui, 2013; Nasrollahzadeh, 2018; Pruna, 2018; Saravanan, 2013; Shejale, 2018; Štengl, 2011; Tang, 2018; Yue, 2010). With the emergence of nanotechnology, a new gate has opened resulting in the drastic increase of interest in photocatalysis and photocatalytically active materials. The ability to tune the size, shape and morphology of nanomaterials help to develop new and advanced physicochemical properties such as increased stability, surface area, optical, magnetic, and catalytic properties have revolutionized the use of photocatalytic active materials (Ashkarran, 2015; Ayati, 2014; Bhanvase, 2017; Giovannetti, 2017; Hatamie, 2015; H. Huang, 2014; Lin, 2012; Lui, 2013; Morales-Torres, 2014; Panwar, 2016; Pruna, 2018; Sanzone, 2018; Saravanan, 2013; Sha, 2014; Shejale, 2018; Shen, 2011; Sugano, 2012; Tang, 2018; Wen, 2011; Y. Yang, 2016; Yue, 2010). However, as a downside, these semiconductor photocatalysts possess a wide bandgap that impedes the electron transportation resulting in the recombination of photo-induced electrons (electron-hole recombination) (Bhanvase, 2017; Ghartavol, 2019; Giovannetti, 2017). This electron-hole recombination occurs so fast that it limits the photocurrent which in turn limits the efficiency of the photocatalyst. Thus, a rising need to improve the existing properties, decelerating the electron-hole recombination and improving the overall efficiency of the material is a topic of interest for the experts worldwide. The commonly used techniques to improve the photocatalytic activity of the generally used semiconductor metal oxides are doping the metal oxide with suitable dopant (Deng, 2011; Wang, 2014), using a composite (Divya, 2017; Kowsari, 2017;

Nanakkal, 2017; Štengl, 2011), and using a composite nanostructure (Ayati, 2014; Chan, 2005; Chen, 2017; Hatamie, 2015; He, 2018; Li, 2018; Morales-Torres, 2014; Panwar, 2016; Qian, 2009; Yun, 2015). A suitable dopant upon introduction into the crystal lattice of the semiconductor will introduce additional energy bands which lowers the bandgap that slows down the fast charge recombination (Bhanvase, 2017; Farhangi, 2011; Štengl, 2011). This will further improve the efficiency of the photocatalyst. Another method to improve the performance of the photocatalyst is to use a nanocomposite. This strategy uses a combination of the photocatalytically active material along with other active components such as carbon nanomaterials, metal oxides etc. to overcome the existing barriers of the photocatalyst (Divya, 2017; Li, 2007; Tang, 2018; Yue, 2010). Another widely studied method to overcome as well as to improve the efficiency of the photocatalyst is to use nanostructures (Liu, 2018; Lui, 2013; Nasrollahzadeh, 2018; Williams, 2008). These nanostructures show enhanced properties that are contributed from the photocatalyst as well as its constituents. Moreover, the nanostructures will be designed in such a way that enables the reactants to utilize the photocatalyst to its full efficiency. Apart from the aforementioned, recent trends to improve the efficiency of the photocatalysis also includes the alteration in synthetic routes such as atomic layer deposition of photocatalytic nanocomposites (Hamza, 2019; Rangarajan, 2018) and to make numerical model analysis(Gao, 2018; Zhao, 2017) of different photocatalytic systems to evaluate the efficiency of various photocatalysts. However, the nanostructured photocatalytic nanocomposites being the hot topic in the current research world, this chapter gives an idea about the recent advancements in nanostructured photocatalytic systems along with the basic principles of photocatalysis, its advantage and disadvantages and some of the commonly discussed applications.

OVERVIEW OF PHOTOCATALYST

Photocatalysis is a combination of both photochemistry and catalysis. The word photocatalysis is of Greek origin meaning '*photo-light*' and '*catalyst-a material that is capable of altering the rate of a reaction without undergoing changes*'. Thus, the photocatalyst is a material that can catalyze a reaction on exposure to the appropriate wavelength of light. The process of utilizing a photocatalyst to alter the rate of reaction is known as photocatalysis. The photocatalyst assisted water splitting on a TiO_2 electrode under UV light was first proposed by Fujishima and Honda in 1972 (Fujishima, 1972). The most important characteristic of a photocatalyst is its ability to concurrently oxidize a donor molecule and reduce an acceptor through the electron-hole pair generated upon excitation.

The process of photocatalysis can generally be divided into two: homogeneous and heterogeneous based on the physical state of the reactants. When both the photocatalyst and reactants are in the same phase such a reaction is called a homogeneous photocatalytic reaction and when both the components are in different phases such a reaction is called heterogeneous photocatalytic reaction. Homogeneous photocatalyst mainly comprises transition metal complexes such as iron, copper, chromium etc. When transition metal complexes are used for photocatalytic reactions, these metal complexes owing to their higher oxidation state generates hydroxyl radicals. The evolved hydroxyl radicals undergo reaction with organic matter that leads to the denaturalization of toxic substances present in the reaction environment. Heterogeneous photocatalysts, on the other hand, are superior to the homogeneous process, as it can be employed for wastewater treatment, where it cuts down many organic pollutants. Some of the common merits of employing heterogeneous photocatalysis are (a) effective waste management, (b) cheap, and (c) mild reaction conditions.

Most commonly semiconductor metal oxides are employed as a heterogeneous photocatalyst for its ability to generate electron-hole pair on exposure to light of appropriate wavelength. For a semiconductor metal oxide to be selected as a photocatalyst, it must possess the following features: it should be a) photoactive b) able to utilize visible and/or near-UV light c) chemically and biologically inert d) stable towards light e) cheap, and f) non-hazardous. Since semiconductor metal oxides satisfy the basic requirement of a photocatalyst, they are widely used as a photocatalyst in different chemical reactions.

Mechanism of Photocatalysis

When a photocatalyst gets incident by the light of appropriate energy, the valence electrons absorb energy from the incident photons and get excited to the conduction band leaving behind a hole in the valence band [Figure 1]. This, in turn, generates an electron-hole pair. The resulting electrons reduce the acceptor as the holes oxidize the donor molecules. Therefore, a photocatalyst can endure simultaneous oxidation and reduction. Similar to other catalytic processes, in photocatalysis, the rate of reaction can be affected by certain parameters including the size, morphology, and surface area of the photocatalyst itself.

Photocatalytically Active Semiconductor Metal Oxides

Semiconductor metal oxides have gained a lot of attention over the past few decades as a photocatalyst owing to its delayed electron-hole recombination which is an important parameter that determines the efficiency of a photocatalyst. The metal

Figure 1. Mechanism of photocatalysis

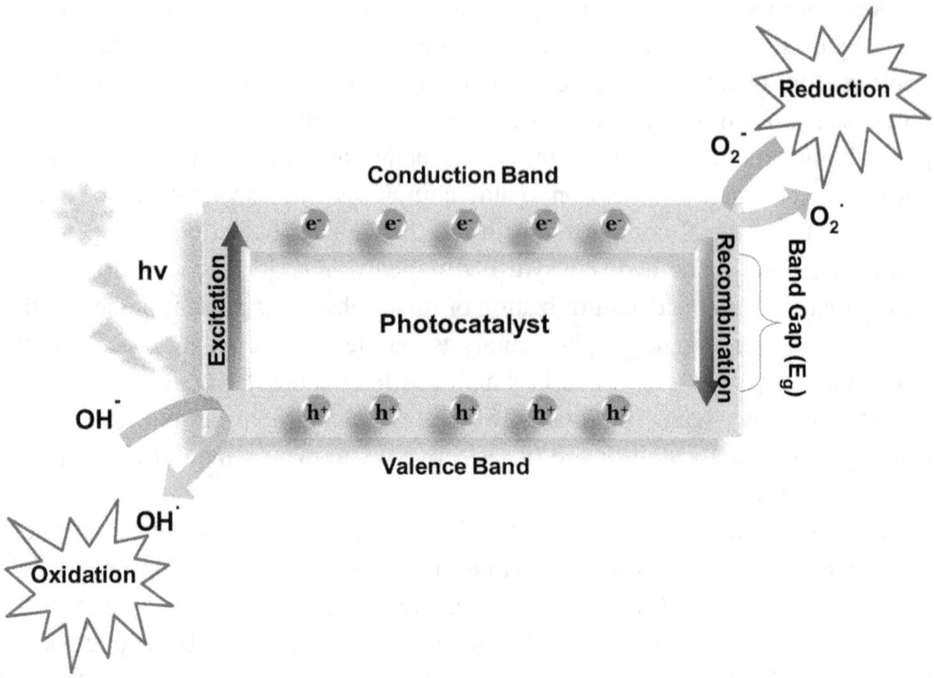

oxides are classified mainly into three: a. binary oxides; b. ternary oxides; and c. quaternary oxides. Binary oxides are again subdivided into 1. titanium dioxide 2. zinc oxide, and 3. other metal oxides such as molybdenum oxide, vanadium oxide, indium oxide, tungsten oxide, and cerium oxide. Ternary and quaternary metal oxides include $ZnFe_2O_4$, $Sr_7Fe_{10}O_{22}$, $Sr_7Fe_{12}O_{19}$, and $Bi_4Nb_xTa_{1-x}O_8I$, $MnCoTiO_2$, Pb_2FeNbO_6 respectively (Ameta, 2018). The aforementioned semiconductor metal oxides have been studied extensively for their effective removal of organic pollutants such as azo dyes, methylene blue etc. Among the many exploited metal oxides, titanium dioxide having a bandgap of 3.32 eV is proved to possess superior ability to break down pollutants because of its increased quantum efficiency, nontoxicity and stability. ZnO and CeO_2 also possess photocatalytic properties comparable to TiO_2 due to their similar bandgap. Although these semiconductor metal oxides are highly active under UV region, they possess very little activity under the visible region (Moma John, 2018). Therefore, various modifications have been done over the years to overcome the limitations of semiconductor metal oxides.

The property enhancement can be done by the formation of localized state just above the valence band or just below the conduction band, employing semiconductors with lower bandgaps, by the formation of colour centre in the bandgap or by surface modification. Thus, the property enhancement of the metal oxide can be done by doping it with a metal or non-metal or co-doping it with various combinations of donor and acceptor materials. Using a combination of photocatalysts and other active materials, i.e. a composite can also enhance the properties of the photocatalyst.

Parameters Affecting Photocatalysis

As a well-known factor, the structure of the catalyst plays a significant role in achieving a higher catalytic activity. For instance, among the three structural phases of TiO_2, anatase phase has superior catalytic properties due to its stability, the position of the conduction band, a higher degree of hydroxylation and absorption power (Khan, 2014). Apart from the catalyst structure, its morphology also plays a key role in determining the activity of the photocatalyst. In the work done by Saravanan et al. (Saravanan, 2013), they stated that spherical ZnO nanoparticles showed a higher efficiency when compared to rod-shaped ZnO particles owing to the larger surface area of the spherical particle.

In the case of a semiconductor photocatalyst, the increase of temperature favours fast electron-hole recombination resulting in decreased catalytic activity. Usually, the system gets the activation energy from photonic activation; an external heat is not required to carry out the reaction. Therefore, the reaction can be conducted at room temperature. Many studies show that the optimum reaction temperature to carryout photocatalysis using TiO_2 falls in the range of 20-80 °C (Saravanan, 2017). The optimum temperature changes depending on the activation energy of the reactants. Another important parameter that affects the rate of the reaction is the pH of the reaction mixture. The studies done in order to understand the effect of pH on the photocatalytic degradation of Reactive Blue 4 using TiO_2 as a catalyst reveals that in acidic conditions i.e. $pH<5$, the dye degradation was lagged due to the high concentration of protons in the system (Reza, 2017; Wang, 2007). On the other hand, in alkaline medium, with a $pH>10$, the increased concentration of hydroxyl radicals degrade the acidic end products formed from the photocatalytic reaction (Reza, 2017; Wang, 2007). Thus an optimum pH should be maintained in order to achieve a good reaction rate. In addition to these, the intensity of incident light and the amount of catalyst also contributes significantly to the rate of reaction.

Applications

The main areas of application of photocatalysts lie in the field of water purification (Ayati, 2014; Ghasemi, 2017; Sanzone, 2018; Shejale, 2018; Yang, 2016) and energy harvesting (Ge, 2016; Li, 2018; Shejale, 2018; Yun, 2015). One of the commonly studied fields of application is the photocatalytic decomposition of air (Zhou, 2016) and water pollutants (Ayati, 2014; Ge, 2016; Nasrollahzadeh, 2018). For instance, doped graphene-TiO_2 nanocomposites show efficient activity on wastewater treatment (Bhanvase, 2017). Another majorly studied application area is its use in solar cells such as in dye-sensitized solar cells (Chou, 2012) and quantum-dot sensitized solar cells (Pawar, 2018) where the semiconductor metal oxide, TiO_2, act as a photoanode. Additionally, photocatalytic systems are also used for solar fuel production by reducing CO_2 (Xie, 2017) and photocatalytic water splitting (Lian, 2015; Mollavali, 2015). Moreover, semiconductors act as sensitizers for photoredox processes due to their electronic structure. Some semiconductors are able to photocatalyse the complete mineralization of many organic pollutants like aromatics (Uribe López, 2019), halo hydrocarbons (Yang, 2011), insecticides (Magalhães, 2017), pesticides (Magalhães, 2017), dyes (Giovannetti, 2017; Yang, 2016), and surfactants (Hatamie, 2015; Liu, 2018; Sugano, 2012). Photocatalytic materials are also used in combination with other active materials in various areas such as decomposition of crude oil (Yue, 2010), photoelectrochemical sensing of Cu^{2+} (Shen, 2011), hydrogen production (Lian, 2015; Liu, 2018) and for antibacterial applications (Anandgaonker, 2015). As discussed before, these semiconductor metal oxides possess many drawbacks that limit their application. Therefore, they are commonly used in combination with other active materials such as potential dopants or as nanocomposites. The applications of such composite systems are discussed in detail in the following sections.

Advantages and Disadvantages of Photocatalysis

The photocatalytic process possesses various advantages such as they are an efficient replacement system for the conventional catalytic methods where an enormous amount of energy is required to carry out the reaction. A photocatalytic system mainly uses renewable and natural energy sources such as solar energy. In contrast to the conventional methods, products evolved from photocatalysis involve non-polluting, non-hazardous products. This technique can be used to treat different pollutants in the wastewater and uses mild reaction conditions which involve lesser reaction times and lower chemical input. Moreover, it generates a negligible amount of waste. However, as the two sides of a coin, the photocatalytic processes also possess some limitations such as interfacial charge transfer, fast electron-hole recombination which adversely affect the efficiency of a photocatalyst when employed in a catalytic process. Therefore, the constant need for enhancing the properties of the photocatalysts is gaining a new realm of interest. To do this utilization of advanced systems such as nanostructured photocatalysts or nanocomposites that possess photocatalytic properties are greatly explored.

NANOSTRUCTURED PHOTOCATALYTIC SYSTEMS-RECENT TRENDS

With the emergence of nanotechnology, many existing processes have triggered to its next level. Recently, heterogeneous photocatalysis is gaining a lot of attention due to its versatility. Heterogeneous photocatalysts fall under the size range of 20-200 nm with increased photosensitivity and high surface area when compared to their bulk counterparts[1]. However, with increased properties, these materials possess an increased bandgap width which reduces their activity in the visible range. In order to improve the existing properties of the photocatalytically active materials, they are used along with other materials that enhance existing properties and impart new functionalities to the catalyst (composites). Lately, much attraction is given to employing such photocatalytically active materials in developing hybrid nanostructures that are multifunctional in nature. Supported structures such as core@ shell (Li, 2018; Yun, 2015), yolk@shell (Dong et al., 2013; He, 2018), and Janus (Huang, 2014; Panwar, 2016) particles that use a combination of photocatalytically active materials and noble metal nanoparticles (Ayati, 2014; He, 2018; Wen, 2011; Yun, 2015; Zhang, 2011) or carbon-based nanomaterials (Ashkarran, 2015; Huang, 2016; Nasrollahzadeh, 2018) have been a recent trend in hybrid photocatalytic nanocomposites. In this chapter, nanostructures that involve TiO_2 and ZnO are given primary importance as they are the most commonly discussed photocatalytic materials. An overall idea about other photocatalytically active materials is also given at the end. A detailed description of all the available materials is beyond the scope of this chapter and will be discussed elsewhere.

TiO_2 Hybrid Nanostructures

Among the many photocatalytically active semiconductor metal oxides, TiO_2 is widely exploited because of its high stability and non-toxicity. However, due to its wide bandgap, these particles are only active in the UV range (Moma John, 2018). It is of high interest for researchers to explore suitable techniques to enhance its activity in the visible range. Carbon-based nanomaterials and noble metal nanoparticles are widely used to enhance TiO_2 due to their large specific surface area, high electron mobility and molecular stability (Liang, 2011).

Lui et al.(Lui, 2013) fabricated graphene (G) wrapped hybrid TiO_2 nanoflowers (G-TiO_2) via a solvothermal reaction [Figure 2]. Their studies revealed that this hybrid structure demonstrated improved efficiency in the photodegradation of methylene blue (MB) under UV light. The increased activity refers to the enhanced charge transport as graphene acts as a conducting channel that suppresses the fast charge recombination. In another work cited by Giovannetti et al.(Giovannetti, 2017), the

Figure 2. (a) and (b) Graphene-TiO_2Nanoflowers with TiO_2 is indicated in red and graphene is indicated in blue; (c) Performance comparison of commercial TiO_2 (P25), TiO_2nanoflower and Graphene-TiO_2nanoflower
[Reprinted from Ref. (Lui, 2013)]

Figure 3. Synthesis route for the preparation of TiO_2/graphene porous nanocomposites (Y. Yang, 2016)

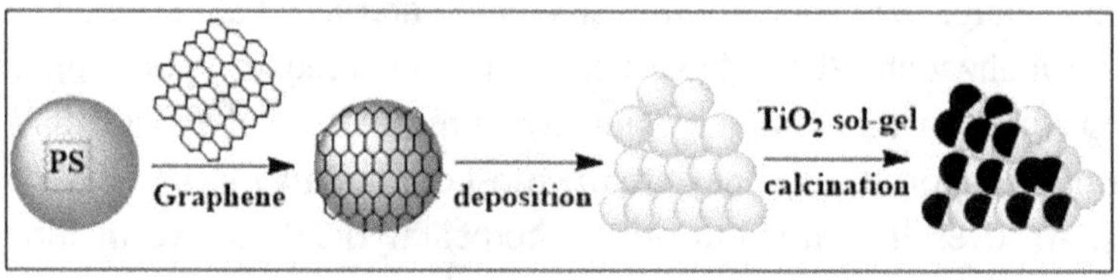

catalytic activity of TiO_2 (anatase) wrapped with graphene oxide (GO) was studied monitoring the degradation of methylene blue (MB) under a xenon lamp.

Yang et al. (Yang, 2016) studied the photocatalytic activity of TiO_2-GO porous composites synthesized via a colloidal crystal template method [Figure 3]. They found that the presence of graphene improved the photocatalytic activity due to its multiple roles. Compared to the TiO_2- graphene interface, the graphene sheets showed an increased affinity towards the adsorption of methylene blue. Moreover, the presence of graphene imparted a better light absorption capacity and an accelerated charge separation with suppression of fast charge recombination to the composite.

In a comparative study done between interpenetrating anatase TiO_2 tablets (IP-TiO_2), overlapping anatase TiO_2 nanosheets (OL-TiO_2), and graphene oxide (GO) supported ultrathin TiO_2 nanosheets (OL-TiO_2/GO), Sha et al.(Sha, 2014) found out that the OL-TiO_2/GO showed improved photocatalytic activity. The presence of GO improved the photocatalytic efficiency by rendering improved specific surface area, accelerating the charge separation and extending the electron life.

Figure 4. Synthetic route for the fabrication of (a) Graphene wrapped TiO_2 yolk@ shells (b) Highly wrapped Graphene TiO_2 yolk@shells
(Ziarati, 2018)

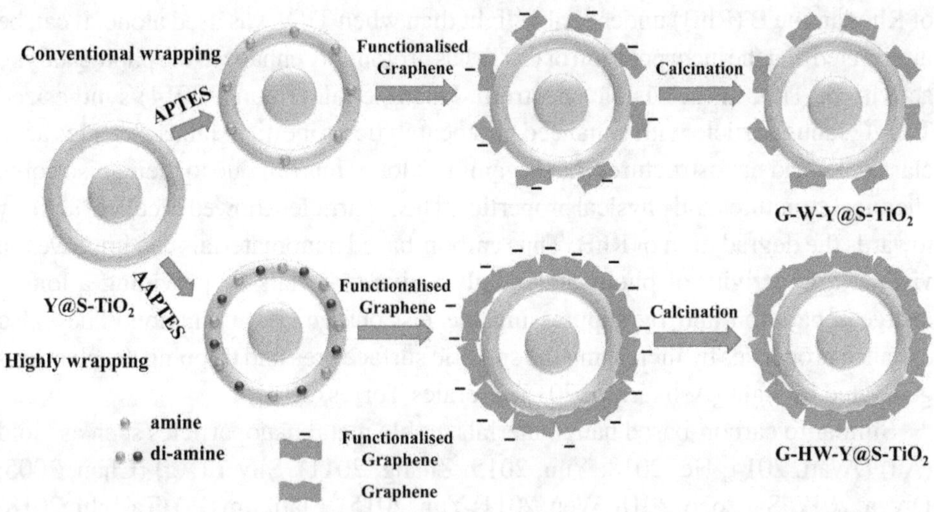

Upon minimizing the defects associated with the 2D sheet structure of graphene, graphene-TiO_2 nanocomposites show improved photocatalytic activity when employed in the reduction of CO_2 under visible light (Liang, 2011). These nanocomposites can be used as a promising candidate in solar fuel production. Ziarati et al. (Ziarati, 2018) developed a yolk@shell structure that comprises of graphene wrapped TiO_2 [Figure 4]. Yolk@shell structures are a special category that falls between core@shell and hollow particles with a void space between the inner core and the outer shell. Such structural assembly can scatter light with a reduced diffusion resistance and increase the photocatalytic efficiency by utilizing the void space which can make the system act as a nanoreactor. The void space provides the necessary space for a reaction to happen as well the shell being porous facilitates the efficient diffusion of reactants and products in and out of the structure. In their work, Ziarati et al. (Ziarati, 2018) demonstrated an effective photocatalytic nanocomposite for the visible light oxidation of aromatic alcohols to its corresponding aldehydes. The structure comprises of TiO_2 nanoparticles wrapped with graphene sheets that not only enhance the visible light absorption of TiO_2 and slows down the charge recombination but also imparts additional properties such as selective photocatalysis of aromatic compounds. The highly conjugated planar structure of graphene facilitates the selective adsorption of aromatic compounds through π-π stacking.

Ashkarranet al. (Ashkarran, 2015) synthesized TiO_2 nanoparticles that are immobilized on carbon nanotubes (CNT) via simple mixing of as-prepared CNTs and TiO_2. This nanocomposite demonstrated improved efficiency in the degradation of Rhodamine B (RhB) under visible light than when TiO_2 was used alone. It can be interpreted that the incorporation of CNTs has profoundly enhanced the photocatalytic activity of TiO_2 in the visible spectrum. Huang et al. (Huang, 2014) synthesized TiO_2/C Janus particle with enhanced photocatalytic properties. Janus particles are a class of hybrid nanostructures that is gaining a lot of interest due to their anisotropic chemical structures and physical properties. These particles showed excellent activity towards the degradation of RhB. Thus carbon-based nanomaterials can improve the visible light activity of photocatalytically active materials by providing a longer charge separation and by suppressing the fast charge recombination. They also enhance properties by increasing the specific surface area and trapping the electrons generated by light (Ashkarran, 2015; Morales-Torres, 2014).

Similar to carbon-based nanomaterials, noble metal nanoparticles such as Gold (Au) (Ayati, 2014; He, 2018; Yun, 2015; Zhang, 2011), Silver (Ag) (Chan, 2005; Divya, 2017; Sanzone, 2018; Wen, 2011; Yun, 2015), Platinum (Pt) (Tasbihi, 2018) and Palladium (Pd) (Liu, 2018) were also investigated for its capability to enhance the photocatalytic efficiency of TiO_2. Noble metal nanoparticle supported TiO_2 show a high photocatalytic activity as the noble metal nanoparticles interact closely with TiO_2 and form Schottky barriers that channels the photogenerated electron from TiO_2 to noble metals which in turn improves the photocatalytic activity (Sugano, 2012; Yu, 2017). Moreover, when Au and Ag are used, their localized surface plasmon resonance (LSPR) enhances the visible light absorption of TiO_2 resulting in an enhanced visible-light photocatalytic activity (Sugano, 2012).

Yu et al. (Yu, 2017) synthesized Au@TiO_2 core-shell particles with truncated wedge-shaped TiO_2 particles for the effective photodegradation of acetaldehyde to P25 in the presence of both UV and visible light. This structure has demonstrated an excellent catalytic activity of selectively localized Au decorated TiO_2 anatase-rutile nanoforests. The nanoforests consisted of nanowires of anatase phase surrounded by radially grown rutile branches. On such assembly, Au NPs were selectively localized. The Au decoration tremendously increased the visible light sensitivity thereby increasing the photocatalytic efficiency due to the effect of LSPR and the electron trapping. In a similar work, Li et al. (Li, 2018) demonstrated the photocatalytic activity of Au, Pd or Pt loaded TiO_2 core-shells. The introduction of a noble metal nanoparticle core to the TiO_2 shell enhances the visible light activity of TiO_2. He et al. (He, 2018) synthesized bio-inspired Au@TiO_2 yolk-shells were which showed excellent photocatalytic activity in the evolution of hydrogen. Tasbihiet al.(Tasbihi, 2018) studied the photocatalytic reduction of CO_2 using Pt/TiO_2 deposited on commercial Al_2O_3 support. This assembly demonstrated enhanced catalytic activity at an optimum loading of Pt. Pt suppress the fast charge recombination of TiO_2 which results in the improvement of the catalyst. Li et al. (Li, 2007) reported the synthesis of highly mesoporous Au/TiO_2 nanocomposites that demonstrated an enhanced photocatalytic activity in the phenol-oxidation and chromium-reduction reactions. They exploited the tunable mesoporous nature of the titania. The porous channels not only provided an increased surface area but also enhanced accessibility for the reactants to reach the respective reaction sites. The tunable mesoporosity makes TiO_2 an important candidate, among the likes such as SiO_2, to be applied in various applications namely separation, catalysis, energy storage, conversion etc. Another type of nanostructure is a Janus particle which is widely studied for its anisotropic properties. For example, TiO_2-SiO_2 Janus nanoparticles synthesized using Pickering emulsion method demonstrated improved photocatalytic activity in the reduction of Solophenyl green dye compared to other TiO_2 nanostructures (Panwar, 2016). From Figure 5 it can be seen that particle labelled as 1 shows a clear Janus morphology whereas the other numbered particles show no nominal anisotropic morphology.

In a work done by Liu et al. (Y. Liu, 2016), they developed a bifunctional membrane that has a three-layer structure with a top layer of TiO_2 nanoparticles, Au NPs as a middle layer and a bottom layer of anodized aluminium oxide (AAO). This composite membrane possesses both photocatalytic as well as photothermal evaporation properties. Thus, it can be used in the solar light assisted water purification. Sanzoneet al. (Sanzone, 2018) developed an Ag/TiO_2 nanocomposite thin film that demonstrated increased photocatalytic activity in the visible range. Wen et al.(Wen, 2011) studied the visible light photocatalytic activity of Ag/TiO_2/graphene nanocomposites. The increased photocatalytic activity in the visible range was due to the strong interaction between TiO_2 nanoparticles and graphene. The LSPR properties of Ag NP also play a significant role in enhancing the visible light activity of the composite. The increase in photocatalytic activity can also be attributed to the strong affinity of graphene towards the adsorption of dye as well as the suppressed charge recombination pace because of the formation of Schottky junction between titania and silver nanoparticles. Another commonly studied noble metal nanoparticle is Palladium (Pd) (Liu, 2018; Zhang, 2011). Liu et al.(P. Liu, 2018) in their work studied the enhanced photocatalytic activity of Pd-TiO_2 hybrid nanostructures in the hydrogen evolution through the breakdown of formic acid.

Other hybrid structures include ternary composites that composed of Au/TiO_2/CNT which possess synergistic properties of all the components. This enhanced the photocatalytic degradation of organic pollutants and their mineralization (Zhang, 2011). Another recent trend in the field of carbon nanomaterials modified TiO_2 nanocomposites is the incorporation of dopants to this system (Bhanvase, 2017). It is expected that the use of dopants will further improve the photocatalytic degradation of different pollutants that cause water pollution. Huang et al. (Huang, 2016) studied

Figure 5. (a) SEM micrograph and (b) EDX mapping of TiO_2-SiO_2 Janus particles [Reprinted from Ref. (Panwar, 2016)]

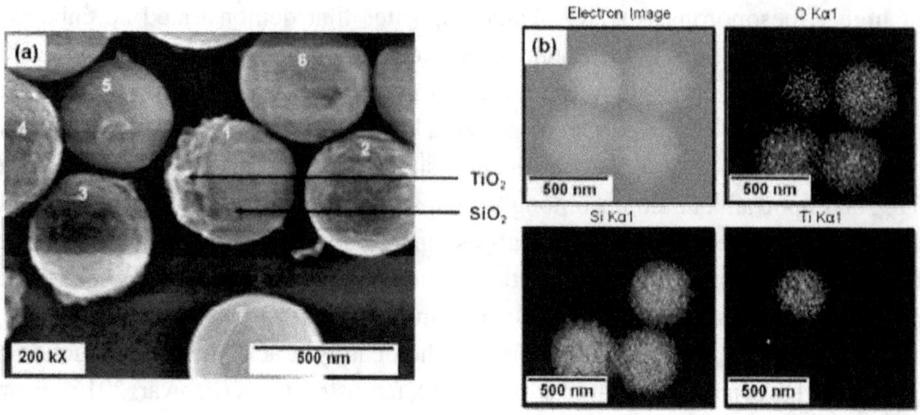

the photocatalytic activity of carbon nitride (CN) and reduced graphene oxide (r-GO) doped TiO_2 that demonstrated notable catalytic activity under visible light towards the degradation of cationic and anionic dyes such as rhodamine B (RhB) and methyl orange (MO) respectively and colourless phenols. They have found that the enhanced activity resulted from a larger specific area and the improved visible light absorption capacity of the hybrid catalyst. A similar system developed by Yue et al. (Yue, 2010) composed of expanded graphite/TiO_2/NiO showed good activity in the decomposition of crude oil. The TiO_2 loaded NiO catalyst was supported on graphite and showed a good ability to absorb heavy oils that are spread in water, which opens up to new possibilities to treat water contaminated with oil. Another study conducted by Lin et al. (Lin, 2012) involved the synthesis of a ternary graphene-TiO_2-Fe_3O_4 nanocomposite that demonstrated enhanced photocatalytic properties towards the degradation of RhB. When compared to the pure TiO_2 and unsupported TiO_2-Fe_3O_4 nanoparticles, this ternary assembly showed an increased efficiency. The presence of graphene suppressed the photodissolution of Fe_3O_4 which renders stability to the nanocomposite thereby improving its reusability. Zhou et al. (Zhou, 2016) developed nanotube arrays of RGO-CeO_2-TiO_2 for the photoelectrodegradation of tetrabromobisphenol A, which is an important brominated retardant that is widely used in PCBs and insulation wires.

ZnO Nanostructures

In addition to TiO_2, ZnO is also a widely studied photocatalyst as they are easily available and when compared to TiO_2, ZnO can absorb a large fraction of solar spectrum. Therefore, ZnO is a good candidate for photocatalytic applications.

However, as a common characteristic of semiconductor metal oxide, ZnO also has a wide bandgap which limits its activity in the visible spectrum. As a result, necessary modifications are required to enhance its properties.

Saravanan et al.(Saravanan, 2013) in their work synthesized a photocatalyst nanocomposite that consists of ZnO/CuO. The photocatalytic activity of the aforementioned composite was tested towards the photocatalytic degradation of MB and MO. Upon comparison with ZnO nanowires, these nanocomposites possess higher catalytic activity. The addition of CuO reduces the bandgap of ZnO which extends the sensitivity of ZnO towards the visible light range. Shen et al. (Shen, 2011) studied the property enhancement of ZnO/CdS nanospheres. Presence of CdS facilitates the quick charge transfer and improves the photocatalytic efficiency of ZnO. López et al.(Uribe López, 2019) investigated ZrO_2 coupled ZnO nanocomposites in the photocatalytic degradation of phenol. Upon comparison to pure ZnO, the ZnO-ZrO_2 nanocomposite showed improved photocatalytic activity with the absence of unwanted intermediate products. ZnO modified with carbon nanomaterials are also widely explored. Work done by Nanakkal et al. (Nanakkal, 2017) studied the modification of ZnO with graphene nanosheet (GN) towards the photodegradation of methylene blue. With respect to pristine ZnO nanostructures, the composite structure of ZnO-GN gave faster degradation rate which reveals the enhancement of properties of ZnO upon coupling with graphene. ZnO nanostructures upon incorporation with activated carbon also showed improved efficiency when used for the degradation of azo dyes such as methyl orange (Nasrollahzadeh, 2018). Graphene oxide (GO) modified ZnO hybrid structures were prepared by Pruna et al.(Pruna, 2018) showcased higher photocatalytic activity due to the synergistic effects of both ZnO and GO. GO provides more specific surface area and suppress the fast charge recombination rendering the nanocomposite a high activity when compared to its individual counterparts.

In addition to the aforementioned, noble metal nanoparticles such as gold nanoparticles (AuNP) can be used to enhance the properties of ZnO in order to utilize them in visible light photocatalysis. Chen et al. studied the photocatalytic activity of Au@ZnO yolk-shells and core-shells(Chen, 2017). They have found that when compared to the core-shells, yolk-shells demonstrated an enhanced catalytic activity. This could be attributed to the peculiarity of the structure a yolk-shell possesses. The porous shell effectively controls the diffusion of reactants into the void space between the core and the shell. Because of the void, masking of the catalytic core by the shell can be avoided. Thus yolk-shells demonstrate an enhanced activity when compared to core-shells.

Another interesting approach in the field of hybrid nanostructures is modified textile for multifunctional applications [Figure 6]. A recent work carried out by Hatamie et al.(Hatamie, 2015) investigated the possibility to use hybrid textile that

Figure 6. Optical Microscopy images of (A) Raw Textile (B, C, D) ZnO NR grown Textile (E) and (F) SEM images of ZnO NR grown Textile at different magnification [Reprinted from Ref. (Hatamie, 2015)]

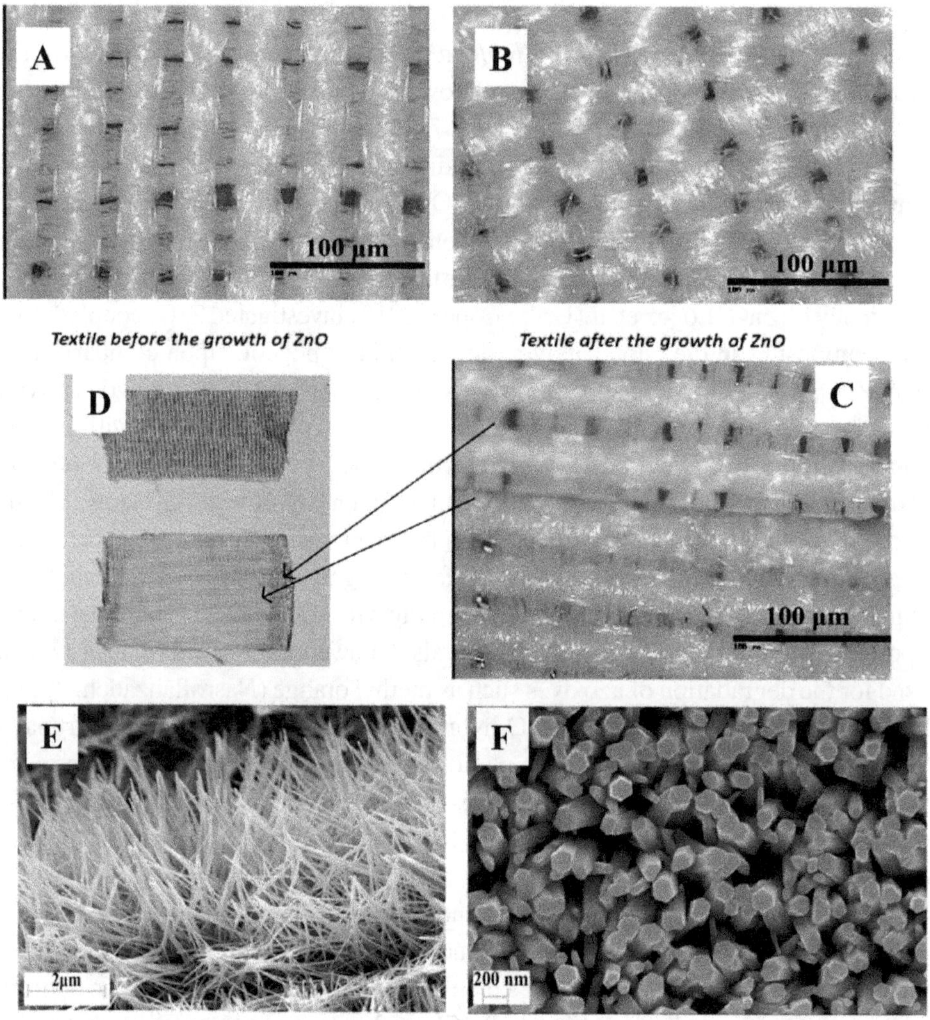

can be used in multifunctional applications such as biosensing, photocatalysis and antibacterial textiles. In this particular work, they successfully grew ZnO nanorods (NR) on a textile substrate which was used for the photocatalytic degradation of organic dyes such as methyl blue and congo red which showed an enhanced catalytic activity than when ZnO NRs are used alone [Figure 7]. Moreover, the same assembly showed a notable response as a flexible biosensor and photocatalytic antibacterial activity.

Figure 7. (A, D) UV-VIS showing the photocatalysis reaction of textile under UV light for (A) MB and (D) CR (254 nm) illumination (until 20 min without UV light); (B, E) Conversion plots after an irradiation time of t (min) (B) MB (E) CR; (C, F) Pseudo-first-order linear plots of ln(C_0/C) versus the irradiation time for the degradation kinetics of (C) MB and (F) CR at an initial dye concentration of 20 mg L^{-1}, catalyst size of 1 cm × 2 cm under UV light irradiation and at pH 7.4.
[Reprinted from Ref. (Hatamie, 2015)]

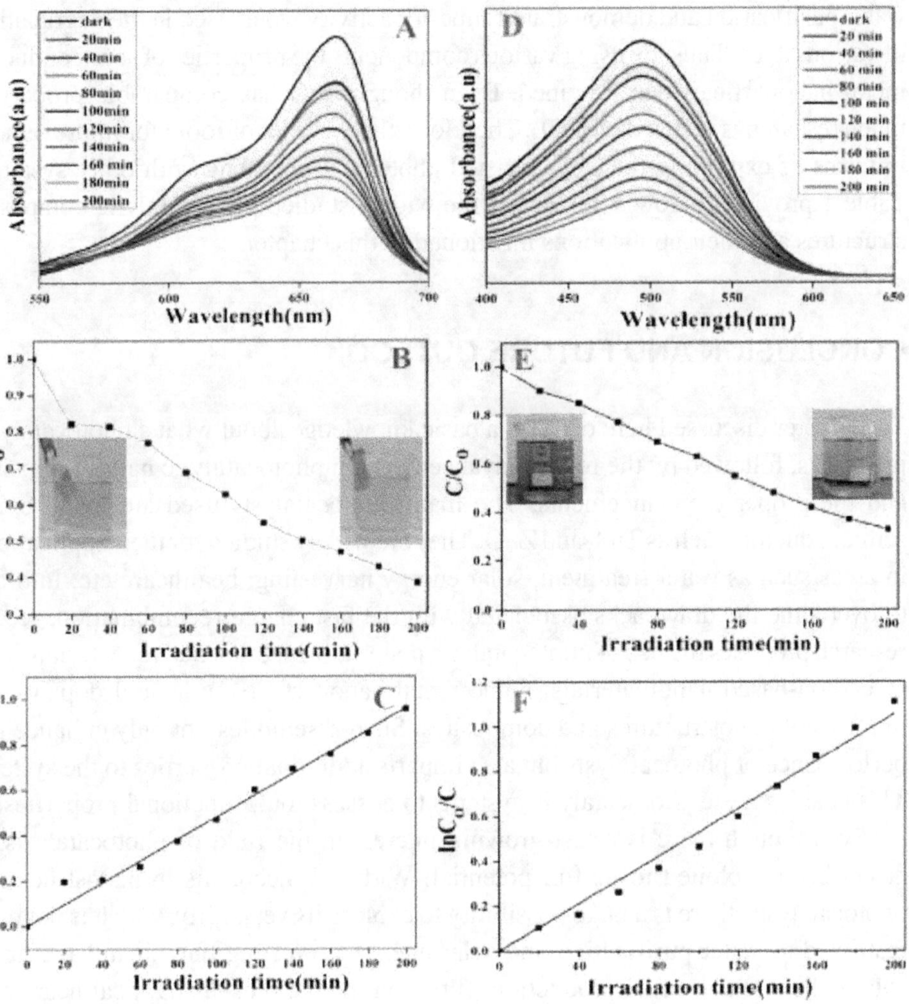

Other hybrid structures involve a combination of three or more materials. For example, the ternary composite of TiO_2/ZnO/RGO/Ag demonstrated enhanced catalytic activity due to the contribution from each component (Divya, 2017). This structure improved the life of charge carriers and due to the presence of co-catalysts,

RGO and Ag, the catalytic efficiency of the photocatalyst was further enhanced. Also, it is gaining lots of interest where a combination of TiO_2 and ZnO is used. Lei et al. (Lei, 2009) developed ZnO nanorods grafted TiO_2 nanotubes. Due to the synergy, they have demonstrated an improved set of properties. As a replacement to noble metal nanoparticles and a carbon material, Cu-based photocatalysts are now gaining interest (Xie, 2017). The Cu modified ZnO shows better photocatalytic activity. Shejaleet al.(Shejale, 2018) investigated the ZnO-TiO_2 nanospheres for water purification and demonstrated superior activity when used in the degradation of various dyes. Thus, by using various components the properties of semiconductor metal nanoparticles can be tuned. Even though ZnO has comparable properties much research is focused on TiO_2. Therefore, there is a lot of room for improvement in terms of exploiting the ZnO assisted photocatalysis along with other systems. Table 1 provided below summarizes the widely studied photocatalytic composite structures and their applications mentioned in this chapter.

CONCLUSION AND FUTURE OUTLOOK

The chapter discussed here delivers a basic knowledge about what a photocatalytic process is, followed by the recent advancements in photocatalytic nanocomposites and the property enhancements. The main photocatalysts used are metal oxide semiconductors such as TiO_2 and ZnO. They are widely studied for their applications in areas such as water treatment, solar energy harvesting, healthcare etc. In order to overcome the drawbacks associated with the fast charge recombination, recent research proposes the use of photocatalytic material with other active materials such as carbon-based nanomaterials, noble metal nanoparticles, potential dopants etc in forms of nanostructures and composites. Such assemblies not only enhance the performance of photocatalysts but also imparts additional properties to the system. This enables these photocatalytic systems to possess multifunctional properties.

Even though there is a fast-growing interest in the field of photocatalysis, it hasn't been exploited to its full potential. With advancements in nanostructured photocatalysts, there is a huge possibility to explore its versatility, which is not only restricted to water purification and solar energy harvesting but expandable up to solar cells and solar fuel production, filtration media, biosensing, healthcare and smart textiles.

Table 1. Summarised list of widely studied photocatalytic composite structures and their applications

Sl. No.	Composite Structure	Application	Ref.
1	G-TiO_2 nanoflowers	Photodegradation of methylene blue under UV light	Chou C. S.2012
2	GO wrapped TiO_2	Photodegradation of methylene blue under Xe lamp	Lei Y. 2009
3	TiO_2-GO porous structure	Photodegradation of methylene blue	Huang M. 2016
4	IP-TiO_2, OL-TiO_2, OL-TiO_2/GO	Photocatalysis in general	Ghasemi S. 2017
5	TiO_2-G (in sheet)	Reduction of CO_2 for solar fuel production	Yue X. 2010
6	TiO_2@G yolk-shells	Visible light oxidation of aromatic alcohols and corresponding aldehydes	Yun J. 2015
7	TiO_2 immobilized on CNT	Degradation of Rhodamine B	Anandgaonker, P. 2015
8	TiO_2/C Janus particles	Degradation of Rhodamine B	Ge, M. (2016)
9	Au@TiO_2 core-shells (with truncated wedge-shaped TiO_2 particles)	Photodegradation of acetaldehyde to P25 under UV and visible light	Zhang, Y. X. 2011
10	Au@TiO_2 yolk-shells	Hydrogen evolution via photocatalytic water splitting	Liu, Y. 2016
11	Pt/TiO_2 deposited on Al_2O_3	Reduction of CO_2	Zhang, N. 2011
12	Mesoporous Au/TiO_2	Phenol-oxidation and chromium-reduction	Morales-Torres, S. 2014
13	TiO_2/SiO_2 Janus particles	Reduction of solophenyl green dye	Hatamie, A. 2015
14	AAO/Au/TiO_2 bi-functional membrane	Solar light assisted water purification	Zhao, X. 2017
15	Pd-TiO_2 hybrid nanostructure	Hydrogen evolution through the breakdown of formic acid	Ayati, A. 2014
16	Carbon nitride and reduced graphene oxide doped TiO_2	Degradation of cationic and anionic dyes as well as colourless phenols	Yu, Y. 2017
17	Graphite/TiO_2/NiO	Decomposition of crude oil	Dong W. 2013
18	Ternary graphene-TiO_2-Fe_3O_4 nanocomposite	Degradation of Rhodamine B	Ghartavol, H. M. 2019
19	Nanotube arrays of RGO-CeO_2-TiO_2	Photoelectrodegradation of tetrabromobisphenol A	Štengl, V. 2011
20	ZnO/CuO nanocomposite, ZnO-GN (graphene nanosheets)	Degradation of anionic and cationic dyes	Chan, S. C. 2005, Lian, Z. 2015
21	ZrO-ZnO nanocomposites	Degradation of phenol	Wang, W. Y. 2007
22	Textile modified with ZnO nanorods	Biosensing, Photocatalysis and antibacterial textiles	Chen, L. 2017
23	ZnO/TiO_2, Cu/ZnO, TiO_2/ZnO/RGO/Ag	Photocatalysis of organic pollutants	Bhanvase, B. A. 2017, Tasbihi, M. 2018, Ziarati, A. 2018

REFERENCES

Ameta, R. (2018). Photocatalysis. In *Advanced Oxidation Processes for Wastewater Treatment*. Emerging Green Chemical Technology. doi:10.1016/B978-0-12-810499-6.00006-1

Anandgaonker, P. (2015). Synthesis of TiO_2 nanoparticles by electrochemical method and their antibacterial application. *Arabian Journal of Chemistry*.

Ashkarran, A. A., Fakhari, M., Hamidinezhad, H., Haddadi, H., & Nourani, M. R. (2015). TiO_2 nanoparticles immobilized on carbon nanotubes for enhanced visible-light photo-induced activity. *Journal of Materials Research and Technology*, *4*(2), 126–132. doi:10.1016/j.jmrt.2014.10.005

Ayati, A., Ahmadpour, A., Bamoharram, F. F., Tanhaei, B., Mänttäri, M., & Sillanpää, M. (2014). A review on catalytic applications of Au/TiO_2 nanoparticles in the removal of water pollutant. *Chemosphere*, *107*, 163–174. doi:10.1016/j.chemosphere.2014.01.040 PMID:24560285

Bhanvase, B. A., Shende, T. P., & Sonawane, S. H. (2017). A review on graphene–TiO_2 and doped graphene–TiO_2 nanocomposite photocatalyst for water and wastewater treatment. *Environmental Technology Reviews*, *6*(1), 1–14. doi:10.1080/21622515.2016.1264489

Chan, S. C., & Barteau, M. A. (2005). Preparation of highly uniform Ag/TiO_2 and Au/TiO_2 supported nanoparticle catalysts by photodeposition. *Langmuir*, *21*(12), 5588–5595. doi:10.1021/la046887k PMID:15924494

Chen, L., Peng, Y., He, F., Liu, G., Huang, Y., Meng, Y., & Guo, R. (2017). Controlled Synthesis and Photocatalytic Performance of Au@ZnO Nanospheres with Core–Shell and Yolk-Shell Structures Assisted by Carbonaceous Layers as Intermediate. *Journal of Nanoscience and Nanotechnology*, *18*(4), 2555–2561. doi:10.1166/jnn.2018.13847 PMID:29442926

Chou, C. S., Chou, F.-C., & Kang, J.-Y. (2012). Preparation of ZnO-coated TiO_2 electrodes using dip coating and their applications in dye-sensitized solar cells. *Powder Technology*, *215–216*, 38–45. doi:10.1016/j.powtec.2011.09.003

Deng, Q. R., Xia, X. H., Guo, M. L., Gao, Y., & Shao, G. (2011). Mn-doped TiO_2 nanopowders with remarkable visible light photocatalytic activity. *Materials Letters*, *65*(13), 2051–2054. doi:10.1016/j.matlet.2011.04.010

Divya, K. S. (2017). A quaternary TiO_2/ZnO/RGO/Ag nanocomposite with enhanced visible light photocatalytic performance. *New Journal of Chemistry*, *41*(14), 6445–6454. doi:10.1039/C7NJ00495H

Dong, W., Zhu, Y., Huang, H., Jiang, L., Zhu, H., Li, C., ... Wang, G. (2013). A performance study of enhanced visible-light-driven photocatalysis and magnetical protein separation of multifunctional yolk–shell nanostructures. *Journal of Materials Chemistry. A, Materials for Energy and Sustainability*, *1*(34), 10030. doi:10.1039/c3ta12272g

Farhangi, N. (2011). Fe doped TiO_2 nanofibers on the surface of graphene sheets for photovoltaics applications. *Photonics North*, *8007*, 800721.

Fujishima, A., & Honda, K. (1972). Electrochemical Photolysis of Water at a Semiconductor Electrode. *Nature*, *238*(5358), 38–40. doi:10.1038/238037a0 PMID:12635268

Gao, F. (2018). Applied Catalysis A, General Stable and highly e ffi cient Cu/TiO_2 nanocomposite photocatalyst prepared through atomic layer deposition. *Applied Catalysis A, General*, *568*(July), 168–175. doi:10.1016/j.apcata.2018.10.006

Ge, M., Cao, C., Huang, J., Li, S., Chen, Z., Zhang, K.-Q., ... Lai, Y. (2016). A review of one-dimensional TiO_2 nanostructured materials for environmental and energy applications. *Journal of Materials Chemistry. A, Materials for Energy and Sustainability*, *4*(18), 6772–6801. doi:10.1039/C5TA09323F

Ghartavol, H. M., Mohammadi, M. R., Afshar, A., & Li, Y. (2019). On the assessment of incorporation of CNT-TiO_2 core-shell structures into nanoparticle TiO_2 photoanodes in dye-sensitized solar cells. *Photochemical & Photobiological Sciences*, *18*(7), 1840–1850. doi:10.1039/C9PP00100J PMID:31204420

Ghasemi, S., Hashemian, S. J., Alamolhoda, A. A., Gocheva, I., & Rahman Setayesh, S. (2017). Plasmon enhanced photocatalytic activity of Au@TiO_2-graphene nanocomposite under visible light for degradation of pollutants. *Materials Research Bulletin*, *87*, 40–47. doi:10.1016/j.materresbull.2016.11.020

Giovannetti, R., Rommozzi, E., Zannotti, M., & D'Amato, C. A. (2017). Recent Advances in Graphene Based TiO_2 Nanocomposites (GTiO2Ns) for Photocatalytic Degradation of Synthetic Dyes. *Catalysts*, *7*(10), 305. doi:10.3390/catal7100305

Hamza, A. (2019). Modelling and optimization of photocatalytic mineralization of acetaminophen using $ZnFe_2O_4$-TiO_2 nanocomposite prepared via microwave assisted combustion method. *J.Mater.Environ.Sci.*, *2508*(3), 290–297.

Hatamie, A., Khan, A., Golabi, M., Turner, A. P. F., Beni, V., Mak, W. C., ... Willander, M. (2015). Zinc Oxide Nanostructure-Modified Textile and Its Application to Biosensing, Photocatalysis, and as Antibacterial Material. *Langmuir*, *31*(39), 10913–10921. doi:10.1021/acs.langmuir.5b02341 PMID:26372851

He, Q., Sun, H., Shang, Y., Tang, Y., She, P., Zeng, S., ... Liu, Z. (2018). Au@ TiO_2 yolk-shell nanostructures for enhanced performance in both photoelectric and photocatalytic solar conversion. *Applied Surface Science*, *441*, 458–465. doi:10.1016/j.apsusc.2018.02.062

Huang, H., Chen, Q., He, J., Zheng, L., Chen, B., & Lin, J. (2014). Synthesis and characterization of TiO_2/C Janus composite particles and its photocatalytic activity for the degradation of rhodamine B. *Colloid & Polymer Science*, *292*(12), 3085–3093. doi:10.100700396-014-3362-1

Huang, M., Yu, J., Hu, Q., Su, W., Fan, M., Li, B., & Dong, L. (2016). Preparation and enhanced photocatalytic activity of carbon nitride/titania(001 vs 101 facets)/ reduced graphene oxide (g-C_3N_4/TiO_2/rGO) hybrids under visible light. *Applied Surface Science*, *389*, 1084–1093. doi:10.1016/j.apsusc.2016.07.180

Khan, M. M., Ansari, S. A., Pradhan, D., Ansari, M. O., Han, D. H., Lee, J., & Cho, M. H. (2014). Band gap engineered TiO_2 nanoparticles for visible light induced photoelectrochemical and photocatalytic studies. *Journal of Materials Chemistry. A, Materials for Energy and Sustainability*, *2*(3), 637–644. doi:10.1039/C3TA14052K

Kowsari, E. (2017). *Carbon-Based Nanocomposites for Visible Light-Induced Photocatalysis*. Academic Press.

Lei, Y., Zhao, G., Liu, M., Zhang, Z., Tong, X., & Cao, T. (2009). Fabrication, characterization, and photoelectrocatalytic application of ZnO nanorods grafted on vertically aligned TiO_2 nanotubes. *The Journal of Physical Chemistry C*, *113*(44), 19067–19076. doi:10.1021/jp9071179

Li, H. (2007a). Mesoporous Au-TiO_2 nanocomposites with enhanced photocatalytic activity. *Journal of the American Chemical Society*, *129*, 67–72. PMID:17381091

Li, H. (2007b). *Mesoporous Au/TiO_2 Nanocomposites with Enhanced Photocatalytic Activity*. Academic Press.

Li, W., Elzatahry, A., Aldhayan, D., & Zhao, D. (2018). Core-shell structured titanium dioxide nanomaterials for solar energy utilization. *Chemical Society Reviews*, *47*(22), 8203–8237. doi:10.1039/C8CS00443A PMID:30137079

Lian, Z. (2015). Plasmonic silver quantum dots coupled with hierarchical TiO_2 nanotube arrays photoelectrodes for efficient visible-light photoelectrocatalytic hydrogen evolution. *Scientific Reports*, *5*(April), 1–10. PMID:26067850

Liang, Y. T., Vijayan, B. K., Gray, K. A., & Hersam, M. C. (2011). Minimizing Graphene Defects Enhances Titania Nanocomposite-Based Photocatalytic Reduction of CO_2 for Improved Solar Fuel Production. *Nano Letters*, *11*(7), 2865–2870. doi:10.1021/nl2012906 PMID:21688817

Lin, Y., Geng, Z., Cai, H., Ma, L., Chen, J., Zeng, J., ... Wang, X. (2012). Ternary graphene-TiO_2-Fe_3O_4 nanocomposite as a recollectable photocatalyst with enhanced durability. *European Journal of Inorganic Chemistry*, *2012*(28), 4439–4444. doi:10.1002/ejic.201200454

Liu, P., Cai, Z., You, Y., Huang, H., Chen, S., Gao, C., ... Xiong, Y. (2018). Surface Modification on Pd-TiO_2 Hybrid Nanostructures towards Highly Efficient H2 Production from Catalytic Formic Acid Decomposition. *Chemistry (Weinheim an der Bergstrasse, Germany)*, *24*(69), 18398–18402. doi:10.1002/chem.201803267 PMID:30102805

Liu, Y., Lou, J., Ni, M., Song, C., Wu, J., Dasgupta, N. P., ... Deng, T. (2016). Bioinspired Bifunctional Membrane for Efficient Clean Water Generation. *ACS Applied Materials & Interfaces*, *8*(1), 772–779. doi:10.1021/acsami.5b09996 PMID:26646606

Lui, G., Liao, J.-Y., Duan, A., Zhang, Z., Fowler, M., & Yu, A. (2013). Graphene-wrapped hierarchical TiO_2 nanoflower composites with enhanced photocatalytic performance. *Journal of Materials Chemistry. A, Materials for Energy and Sustainability*, *1*(39), 12255–12262. doi:10.1039/c3ta12329d

Magalhães, P. (2017). Titanium dioxide photocatalysis: Fundamentals and application on photoinactivation. *Reviews on Advanced Materials Science*, *51*(2), 91–129.

Mollavali, M., Falamaki, C., & Rohani, S. (2015). Preparation of multiple-doped TiO_2 nanotube arrays with nitrogen, carbon and nickel with enhanced visible light photoelectrochemical activity via single-step anodization. *International Journal of Hydrogen Energy*, *40*(36), 12239–12252. doi:10.1016/j.ijhydene.2015.07.069

Moma, J. (2018). Modified Titanium Dioxide for Photocatalytic - Photocatalysts - Applications and Attributes. *IntechOpen*, 37–56.

Morales-Torres, S. (2014, July). Nanostructured carbon–TiO_2 photocatalysts for water purification : an overview Fotocatalizadores carbón-TiO_2 para purificación de aguas : revisión. *Grupo Español Carbón*, 9–14.

Nanakkal, A. R., & Alexander, L. K. (2017). Photocatalytic Activity of Graphene/ZnO Nanocomposite Fabricated by Two-step Electrochemical Route. *Journal of Chemical Sciences*, *129*(1), 95–102. doi:10.100712039-016-1206-x

Nasrollahzadeh, M. S., Hadavifar, M., Ghasemi, S. S., & Arab Chamjangali, M. (2018). Synthesis of ZnO nanostructure using activated carbon for photocatalytic degradation of methyl orange from aqueous solutions. *Applied Water Science*, *8*(4), 1–12. doi:10.100713201-018-0750-6

Panwar, K., Jassal, M., & Agrawal, A. K. (2016). TiO_2-SiO_2Janus particles with highly enhanced photocatalytic activity. *RSC Advances*, *6*(95), 92754–92764. doi:10.1039/C6RA12378C

Pawar, S. A., Jeong, J. P., Patil, D. S., More, V. M., Lee, R. S., Shin, J. C., & Choi, W. J. (2018). Quantum Dot Sensitized Solar Cells Based on TiO_2/$AgInS_2$. *Journal of the Korean Physical Society*, *72*(10), 1188–1193. doi:10.3938/jkps.72.1188

Pruna, A., Wu, Z., Zapien, J. A., Li, Y. Y., & Ruotolo, A. (2018). Enhanced photocatalytic performance of ZnO nanostructures by electrochemical hybridization with graphene oxide. *Applied Surface Science*, *441*, 936–944. doi:10.1016/j.apsusc.2018.02.117

Qian, J., Liu, P., Xiao, Y., Jiang, Y., Cao, Y., Ai, X., & Yang, H. (2009). TiO_2-coated multilayered SnO2 hollow microspheres for dye-sensitized solar cells. *Advanced Materials*, *21*(36), 3663–3667. doi:10.1002/adma.200900525

Rangarajan, G. (2018, June). Synthesis of MgO / TiO_2 Nanocomposite and Its Application in Photocatalytic Dye Degradation. *International Journal of Chemical Reactor Engineering*.

Reza, K. M., Kurny, A. S. W., & Gulshan, F. (2017). Parameters affecting the photocatalytic degradation of dyes using TiO2: A review. *Applied Water Science*, *7*(4), 1569–1578. doi:10.100713201-015-0367-y

Sanzone, G., Zimbone, M., Cacciato, G., Ruffino, F., Carles, R., Privitera, V., & Grimaldi, M. G. (2018). Ag/TiO_2 nanocomposite for visible light-driven photocatalysis. *Superlattices and Microstructures*, *123*, 394–402. doi:10.1016/j.spmi.2018.09.028

Saravanan, R., Karthikeyan, S., Gupta, V. K., Sekaran, G., Narayanan, V., & Stephen, A. (2013). Enhanced photocatalytic activity of ZnO/CuO nanocomposite for the degradation of textile dye on visible light illumination. *Materials Science and Engineering C*, *33*(1), 91–98. doi:10.1016/j.msec.2012.08.011 PMID:25428048

Saravanan, R. (2017). Basic Principles, Mechanism, and Challenges of Photocatalysis. Springer.

Sha, J., Zhao, N., Liu, E., Shi, C., He, C., & Li, J. (2014). In situ synthesis of ultrathin 2-D TiO_2 with high energy facets on graphene oxide for enhancing photocatalytic activity. *Carbon*, *68*, 352–359. doi:10.1016/j.carbon.2013.10.087

Shejale, K. P., Laishram, D., Gupta, R., & Sharma, R. K. (2018). Engineered ZnO-TiO_2 Nanospheres for High Performing Membrane Assimilated Photocatalytic Water Remediation and Energy Harvesting. *ChemistrySelect*, *3*(25), 7291–7301. doi:10.1002lct.201800988

Shen, Q., Zhao, X., Zhou, S., Hou, W., & Zhu, J.-J. (2011). ZnO/CdS hierarchical nanospheres for photoelectrochemical sensing of Cu2+. *The Journal of Physical Chemistry C*, *115*(36), 17958–17964. doi:10.1021/jp203868t

Štengl, V., Popelková, D., & Vláčil, P. (2011). TiO_2-graphene nanocomposite as high performace photocatalysts. *The Journal of Physical Chemistry C*, *115*(51), 25209–25218. doi:10.1021/jp207515z

Sugano, Y. (2012). Gold Nanoparticles Located at the Interface of Anatase/Rutile TiO_2 Particles as Active Plasmonic Photocatalysts for Aerobic Oxidation. *Journal of the American Chemical Society*, *134*(14), 6309–6315. doi:10.1021/ja2120647 PMID:22440019

Tang, B., Chen, H., Peng, H., Wang, Z., & Huang, W. (2018). Graphene Modified TiO_2 Composite Photocatalysts: Mechanism, Progress and Perspective. *Nanomaterials (Basel, Switzerland)*, *8*(2), 105. doi:10.3390/nano8020105 PMID:29439545

Tasbihi, M. (2018). Pt/TiO_2 photocatalysts deposited on commercial support for photocatalytic reduction of CO_2. *Journal of Photochemistry & Photobiology, A: Chemistry*.

Uribe López, M. C., Alvarez Lemus, M. A., Hidalgo, M. C., López González, R., Quintana Owen, P., Oros-Ruiz, S., ... Acosta, J. (2019). Synthesis and Characterization of ZnO-ZrO_2 Nanocomposites for Photocatalytic Degradation and Mineralization of Phenol. *Journal of Nanomaterials*, *2019*, 1–12. doi:10.1155/2019/1015876

Wang, L., Fan, J., Cao, Z., Zheng, Y., Yao, Z., Shao, G., & Hu, J. (2014). Fabrication of predominantly Mn^{4+}-Doped TiO_2 nanoparticles under equilibrium conditions and their application as visible-light photocatalyts. *Chemistry, an Asian Journal*, *9*(7), 1904–1912. doi:10.1002/asia.201402114 PMID:24839127

Wang, W. Y., & Ku, Y. (2007). Effect of solution pH on the adsorption and photocatalytic reaction behaviors of dyes using TiO_2 and Nafion-coated TiO_2. *Colloids and Surfaces. A, Physicochemical and Engineering Aspects*, *302*(1–3), 261–268. doi:10.1016/j.colsurfa.2007.02.037

Wen, Y., Ding, H., & Shan, Y. (2011). Preparation and visible light photocatalytic activity of Ag/TiO_2/graphene nanocomposite. *Nanoscale*, *3*(10), 4411–4417. doi:10.1039/c1nr10604j PMID:21909581

Williams, G., Seger, B., & Kamat, P. V. (2008). UV-Assisted Photocatalytic Reduction of Graphene Oxide. *ACS Nano*, *2*(7), 1487–1491. doi:10.1021/nn800251f PMID:19206319

Xie, H., Wang, J., Ithisuphalap, K., Wu, G., & Li, Q. (2017). Recent advances in Cu-based nanocomposite photocatalysts for CO_2 conversion to solar fuels. *Journal of Energy Chemistry*, *26*(6), 1039–1049. doi:10.1016/j.jechem.2017.10.025

Yang, X., Li, Y., Lu, A., Yan, Y., Wang, C., & Wong, P.-K. (2011). Photocatalytic reduction of carbon tetrachloride by natural sphalerite under visible light irradiation. *Solar Energy Materials and Solar Cells*, *95*(7), 1915–1921. doi:10.1016/j.solmat.2011.02.020

Yang, Y., Xu, L., Wang, H., Wang, W., & Zhang, L. (2016). TiO2/graphene porous composite and its photocatalytic degradation of methylene blue. *Materials & Design*, *108*, 632–639. doi:10.1016/j.matdes.2016.06.104

Yu, Y. (2017). UV and visible light photocatalytic activity of Au/TiO_2 nanoforests with Anatase/Rutile phase junctions and controlled Au locations. *Scientific Reports*, *7*, 1–13.

Yue, X., Zhang, R., Zhang, F., & Wang, L. (2010). Decomposition of crude oil absorbed into expanded graphite/TiO_2/NiO composites. *Desalination*, *252*(1–3), 163–166. doi:10.1016/j.desal.2009.10.007

Yun, J., Hwang, S. H., & Jang, J. (2015). Fabrication of Au@Ag core/shell nanoparticles decorated TiO_2 hollow structure for efficient light-harvesting in dye-sensitized solar cells. *ACS Applied Materials & Interfaces*, *7*(3), 2055–2063. doi:10.1021/am508065n PMID:25562329

Zhang, N., Liu, S., Fu, X., & Xu, Y.-J. (2011). Synthesis of M@TiO_2 (M = Au, Pd, Pt) core-shell nanocomposites with tunable photoreactivity. *The Journal of Physical Chemistry C*, *115*(18), 9136–9145. doi:10.1021/jp2009989

Zhang, Y. X., Gao, B., Li Puma, G., Ray, A. K., & Zeng, H. C. (2011). Self-Assembled Au/TiO_2/CNTs Ternary Nanocomposites for Photocatalytic Applications. *Science of Advanced Materials*, *2*(4), 503–513. doi:10.1166am.2010.1116

Zhao, X. (2017). Photocatalytic Properties of Co_3O_4 -Coated TiO_2 Powders Prepared by Plasma- Enhanced Atomic Layer Deposition. *Nanoscale Research Letters*, *12*(497). PMID:28815483

Zhou, Q., Xing, A., Zhao, D., & Zhao, K. (2016). Tetrabromobisphenol A photoelectrocatalytic degradation using reduced graphene oxide and cerium dioxide comodified TiO_2 nanotube arrays as electrode under visible light. *Chemosphere*, *165*, 268–276. doi:10.1016/j.chemosphere.2016.08.143 PMID:27657819

Ziarati, A. (2018). Engineered Bi-Functional Hydrophilic/Hydrophobic Yolk@ Shell Architectures: A Rational Strategy for Non-Time Dependent Ultra Selective Photocatalytic Oxidation. *Applied Catalysis B: Environmental*, *18*, 1–19.

Chapter 7
Recent Developments of Epoxy Nanocomposites Used for Aerospace and Automotive Application

Sudheer Kumar
https://orcid.org/0000-0002-8944-0728
School for Advanced Research in Polymers, Central Institute of Plastics Engineering and Technology, Bhubaneswar, India

Sukhila Krishnan
Sahrdaya College of Engineering and Technology, India

Sushanta Kumar Samal
School for Advanced Research in Polymers, Central Institute of Plastics Engineering and Technology, Bhubaneswar, India

ABSTRACT

Epoxy resins are widely utilized engineering thermosetting polymers for industrial applications such as aerospace and automotive fields due to their higher mechanical, thermal, and chemical resistance. Recently, polymer nanocomposites have attracted huge attention both in academics and industry because they demonstrated the tremendous enhancement in material properties compared with a neat polymer or traditional micro and macro composites. Traditional composites generally require a high content (>10%) of the inorganic fillers to bestow the desired mechanical properties. Higher filler content raises the density of the new product, thereby reducing the properties through fragile interfacial interaction among filler and the organic matrix. Furthermore, enhancing filler content makes processability very complicated. However, nanocomposites exhibit improved thermomechanical properties even with a small amount of nanoparticles (≤5%). The chapter provides information about the application of polymer nanocomposites (i.e., aerospace and automotive industries).

DOI: 10.4018/978-1-7998-1530-3.ch007

INTRODUCTION

Epoxy resin is one of the most extensively employed traditional thermosetting material, which has wide industrial applications including coatings, (Shi et al., 2009; Acebo et al., 2014) adhesives, (Hsiao et al., 2003; Zhu et al., 2010) electronics, (Tuncer et al., 2006) marine, (Tian et al., 2014; Wang et al., 2014) aerospace (Toldy et al., 2011; Gu et al., 2013) and automobiles due to its higher mechanical, thermal, properties, solvent resistance and good insulation properties (Jyotishkumar et al., 2009; Gu et al., 2013). Conversely, after incorporation of nanofiller like CNT, (Qing et al., 2014; Cui et al., 2013) GO (Dreyer et al. 2010; He et al., 2010) SiO_2, (Qi et al., Kumar et al., 2017) (TiO_2 (Chatterjee et al., 2008; Kumar et al., 2016) into the epoxy matrix, the performance of the epoxy nanocomposites was further enhanced. Polymer nanocomposites (PNCs) are potential substitutes to microcomposites and monolithic owing to their excellent properties. Nevertheless, the fabrication method of the polymer nanocomposites faces a lot of challenges due to the control of elemental composition and stoichiometry ratios in the nanophase (Khan et al. 2016).

PNCs are materials to which nano-range filler parts are incorporated in order to enhance the characteristic performance of the final materials (Asmatulu et al., 2015). PNCs have two different components of phase constituting dissimilar physical and chemical properties and are segregated via a different interface. Their different characteristic properties are not exhibited through one of its constituents. The constituent which is commonly available in the large quantity is called polymer matrix. Whereas, other constituents incorporated into the matrix which enhances the mechanical properties of the PNCs are called reinforcement (i.e., nanomaterials) (Sunil et al., 2019).

Usually, nanocomposites demonstrated anisotropy characters (directionally dependent) due to the different properties of constituents and inhomogeneous dispersion of the nanomaterials. PNCs exhibits more benefits as compared to traditional composites such as; (i) The enhancement in properties of the polymer matrix in nanocomposites can be obtained via the incorporation of small quantities of nanomaterial in contrast to traditional composites that need higher concentrations of microparticles for the enhancement of the properties, (ii) Owing to the inclusion of the tiny percentage of nanomaterials, PNCs are much lighter in weight than traditional composites and (iii) Nanomaterials having size-related properties improves mechanical, thermal, chemical and electrical properties to a much greater range compared to traditional composites (Jannapu et al., 2010).

PNCs are usually described as the arrangement of polymer matrix and filler at minimum one dimension in the nanometer range. The nanofiller can be divvied into three parts such as one dimensional (e.g., MMT nanoclay), two-dimensional (e.g., carbon nanotube, nanofibers, nanowires etc.,), three-dimensional (e.g., silica nanoparticles, nanowhiskers etc.) (Tiwari et al., 2012; Khan et al., 2016). Polymer nanocomposites are notoriously for its remarkable mechanical properties like higher elastic modulus, enhanced strength, flame retardancy and barrier resistance with the inclusion of very little quantity of nano (5%) particles. This is attributed to the extremely huge surface area of interaction among polymer matrix and nanofiller. Figure 1 demonstrated that the overall application of the polymer nanocomposites.

The present chapter focused to deliberate the contribution of nanomaterials towards the extensive enhancement of characteristic properties of the polymer nanocomposites for aerospace and automotive applications after the surface treatment of nanomaterials and provides the additional functionality. Therefore, nanotechnology bestows the platform to synthesize high-performance polymeric nanocomposites with outstanding properties. Additional, to discuss of PNCs influence via the surface treatment of nanomaterials, preparation techniques. Therefore, PNCs have been extensively used for overall applications from sensing and actuation, information industry, defense system, biomedical and electronic industry etc., as shown in Figure 1 (Tao et al., 2013; Wang et al., Kao et al). But in this chapter, the main focus will be put on the PNCs for aerospace and automotive industry. Further, Table 1 also shows that the nanomaterials applications in aerospace and automotive.

PREPARATION METHODS OF EPOXY POLYMER NANOCOMPOSITES

There are numerous methods for the preparation of polymer nanocomposites. These incorporate in solution casting; melt intercalation and in situ polymerization

Solution Casting

Solution casting is one of the most simple, easy and effective generally used techniques to fabricate uniform dispersed polymer nanocomposites (PNCs) (Mital et al.). In this process, the nanomaterials are dispersed in an appropriate solvent in which the polymer is soluble. After that, the solvent was removed via evaporation at ambient temperature and the polymer nanocomposites sheet was developed as shown in Figure 2 (Singh et al., 2014; Chandrasekaran et al., 2013).

Melt Blending

Melt blending is another cost-effective, the eco-friendly process for the synthesis of polymer nanocomposites, generally thermoplastic polymer nanocomposites as represented in Figure 3 (Dennis et al,, Ge et al., 2017). No solvent is required for the

Figure 1. Polymer nanocomposites application in various fields

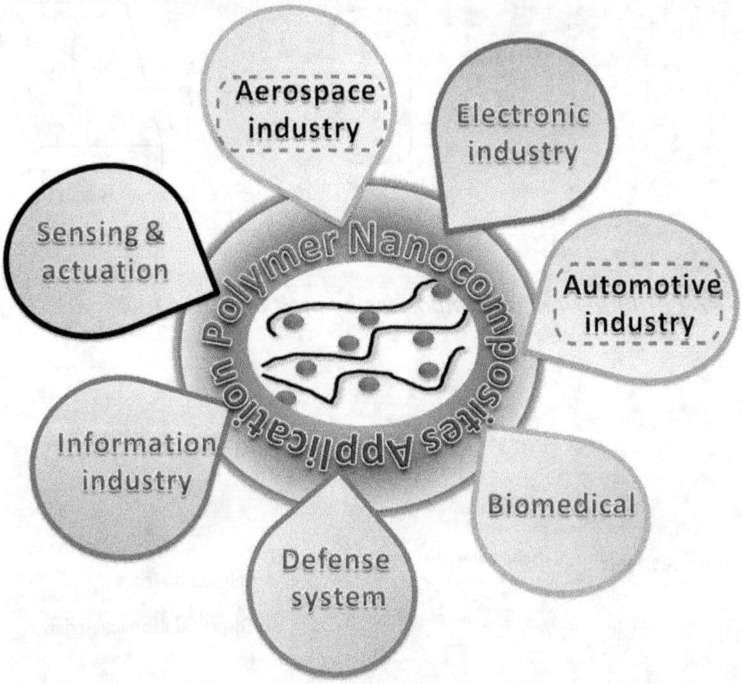

Table 1. Nanofiller used in the aerospace and automotive applications

Nanofiller	Area	Applications
Carbon nanotube (CNT)	Aerospace	• Fuselage and wings.
	Automotive	• External body. • Timing belt covers in automotive. • Scratch-resistant coatings in automotive and aircraft frame body.
Graphene (G) & Graphene oxide (GO)	Aerospace	• Bestow electrostatic discharge and electromagnetic interference shielding parts plus weight saving.
	Automotive	• Electrostatic spray painting in body components.
Nanocaly	Aerospace	• Cryogenic storage system.
	Automotive	• Manufacturing engine, timing belt covers structural seatbacks, and fuel system parts.
Titanium oxide (TiO_2)	Aerospace	• Coating application in aerospace.
	Automotive	• Automotive coatings are among the most demanding applications for TiO_2 pigments. • Ti-Pure™ grades are available for automotive coating applications, offering both Original Equipment Manufacturers (OEMs)

Figure 2. Preparation of nanocomposites via solution casting process

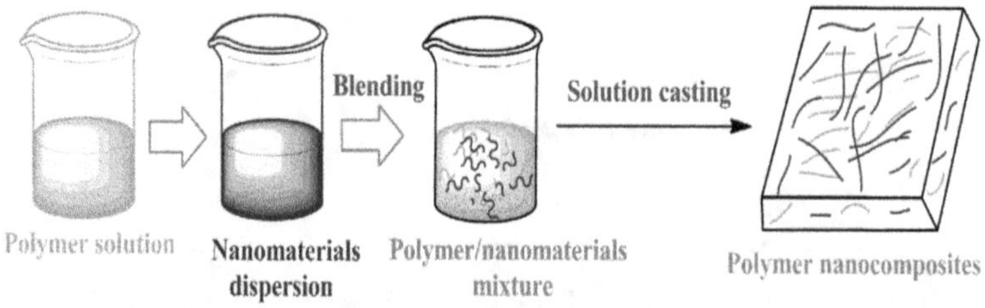

Figure 3. Preparation of nanocomposites by melt blending method

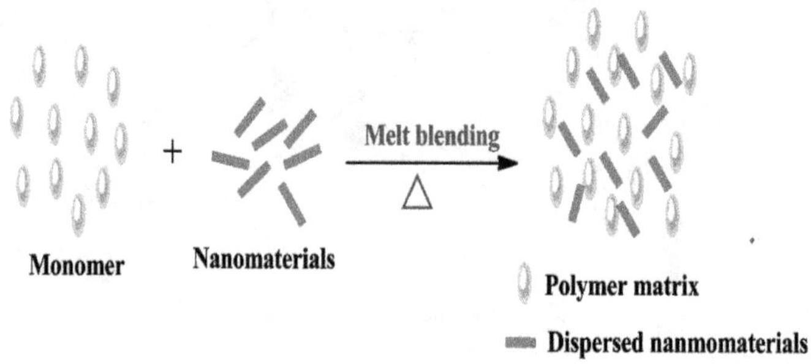

synthesis of PNCs and it is considered to be one of the useful and well-organized methods in an industrial point of view for huge scale productions. In the last few years, lots of industries have employed this method for the preparation of polymer nanocomposites (Utracki et al., 2004).

In-Situ Polymerization

Nanomaterials/polymer nanocomposites are generally prepared using in-situ polymerization method as depicted in Figure 4 (Bose et al., 2010; Wu et al., 2012). In-situ polymerization method provides a strong interaction amongst the polymer matrix and the nanofiller. The monomer and nanomaterials are first dissolved in a general solvent and were undergone ultrasonication to obtain a homogeneous dispersion and polymerizations were initiated via suitable initiator. It is noticed that the polymer nanocomposites prepared via this method demonstrated better mechanical properties compared to the solution and melt blending techniques (Choi et al., 2010; Chee et al., 2015).

Figure 4. Preparation of nanocomposites by an in-situ polymerization method

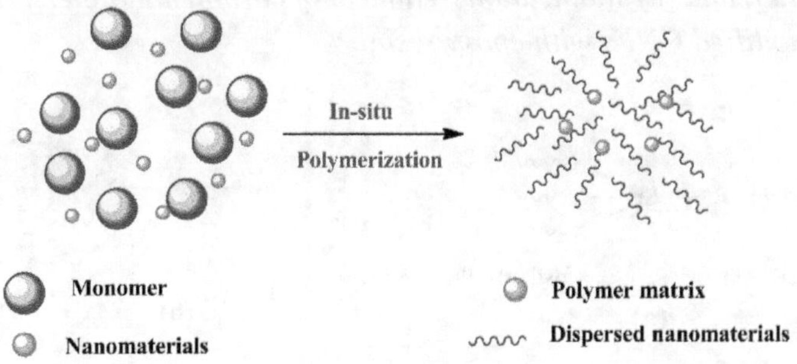

PROBLEMS AND SOLUTION FOR THE SYNTHESIS OF EPOXY NANOCOMPOSITES

Commonly, due to the higher surface to volume ratio, nanomaterials are abundant facile to be placed continuously and attract each other to form agglomeration. Consequently, for the duration synthesis of PNCs, two major problems endure for achieving the strengthened polymer nanocomposites. For example, nanofiller dispersion and interfacial interaction between polymer and nanofiller (Chan et al., 2002; Wang et al., 2011; Guo et al., 2007). The polymer nanofiller interfaces provide a crack starting point which declines the mechanical properties of the nanocomposites. Generally, nanoparticles dispersion can be achieved via surface modification with particular surfactants, coupling agents, polymer (Guo et al. 2006) and by successful stirring. Various methods for physical stirring have been investigated to disperse nanofiller such as mechanical, magnetic stirring, beads milling, shear mixing, three roller mill and ultrasonication (Potts et al., 2011).

Silane Treatment

Silane is generally used as a coupling agent, it possesses two carboxylic groups. One functional group can make a bond with nanofiller while, another can connect to the matrix as depicted in Figure 5 (a,b). However, the coupling agents are generally employed to bestow a stable bonding network among the nanofiller and the matrix (Zekri et al., 2007). Further, the applied load also transfer from fragile polymer to the stronger nanofiller which results in improved reinforcement performance and bestow the longer service life of polymer matrix (Gu et al., 2013; Gu et al., 2009). Figure 5c demonstrates the example of functionalization on the surface of carbon nanofibers (CNFs) by silanization and by using a 3-aminopropyl-triethoxysilane (APTES) as an amine-based coupling agent.

Figure 5. Chemical structure of (a) coupling agent and (b) network among the matrix, inorganic filler, and (c) silanization treatment of carbon nanofibers, curing system of APTES modified CNFs with epoxy resin.

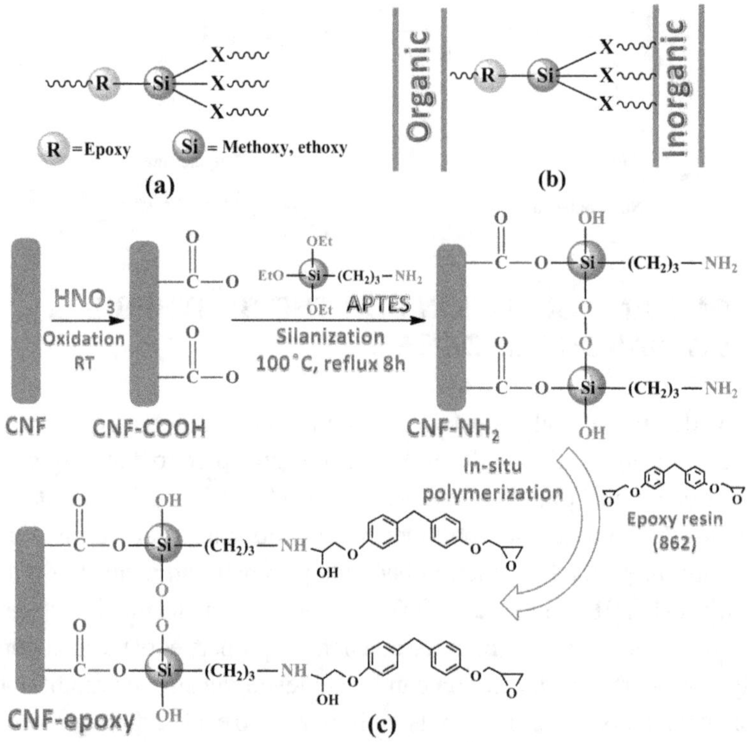

The silanization process favour to CNF distribution and enhance the interfacial interaction among the CNFs and the epoxy resin by the formation of chemical bonding as shown in Figure 5c. Initially, the alkoxy group of APTES reacts with the CNFs surface carboxylic group (-COOH) and an amine group on APTES was reacted with the oxirane group of the epoxy resin (Zhu et al., 2010). Further, another type of silane like 3-glycidoxypropyltrimethoxysilane (GPTMS) can also be employed as a coupling agent to enhance the dispersion properties of the carbon nanotubes (CNT) within the epoxy matrix (Ma et al., 2007).

Further, Figure 6 demonstrates a functionalized surface of epoxy resin using a 3-aminopropyl-triethoxysilane (APTES) as an amine-based coupling agent and forming a more compatible network which, generates lower phase separation among the organic and inorganic phases and enhanced the mechanical, thermal properties of the cured products (Kumae et al., 2002; Ji et al., 2006; Seraj et al., 2014).

Figure 6. Synthetic route modification of epoxy resin with aminopropyltriethoxy silane (APTES).

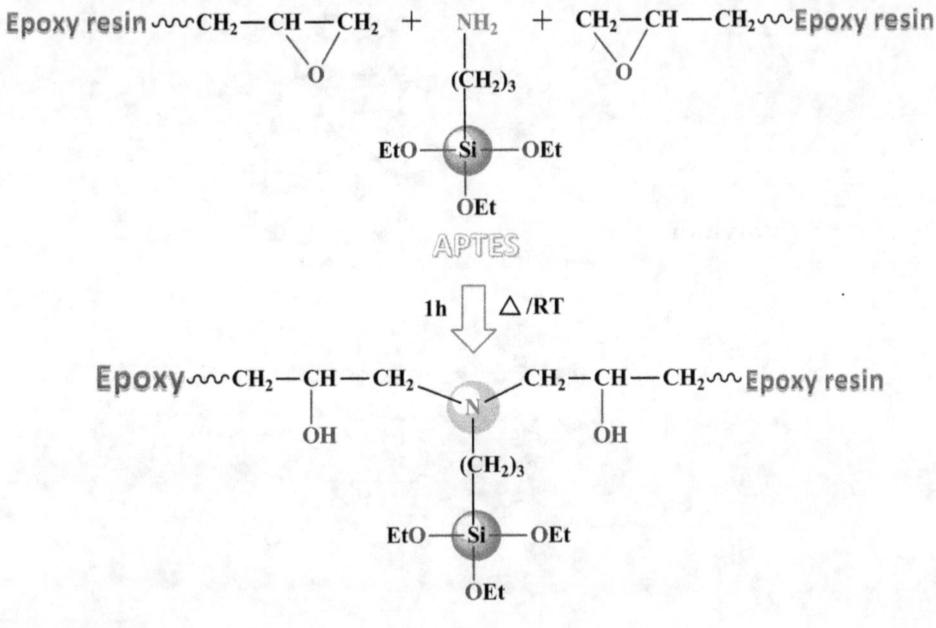

Polymer Grafting

In the preparation of nanofiller (e.g., CNT, graphene, clay and TiO_2 based nanocomposites, the interface of the nanofiller materials and the polymer matrix plays an important character in deciding the final characteristic properties of the prepared product as illustrated in Figure 7. Consequently, polymer grafting method is generally used to incorporate functional groups on the surface of the above mention nanofiller and performs as a compatibilizing or anchoring agent to enhance the interfacial adhesion of the immiscible parts and therefore, raise the compatibility and dispersion of the nanofiller within the matrix. Usually, various kinds of monomer have been employed in the grafting method (Rodriguez et al., 2012; Tantis et al., 2012; Hung et al., 2014; Thangavel et al., 2014; Pham et al., 2012). The selection of the monomer based exclusively on the polarity and kinds of functional groups existing on the surface of the polymer matrix.

Another process for modification of the surface of inorganic and organic materials is based on the grafting of polymeric materials to the substrate and improving the chemical functionality and changes the surface topography of the initial materials (Kango et al., 2013). The polymer grafted inorganic nanoparticles are observed to be organic-inorganic nanocomposites particles. As the monomer

Figure 7. Polymer grafting representation with the carbon nanotube

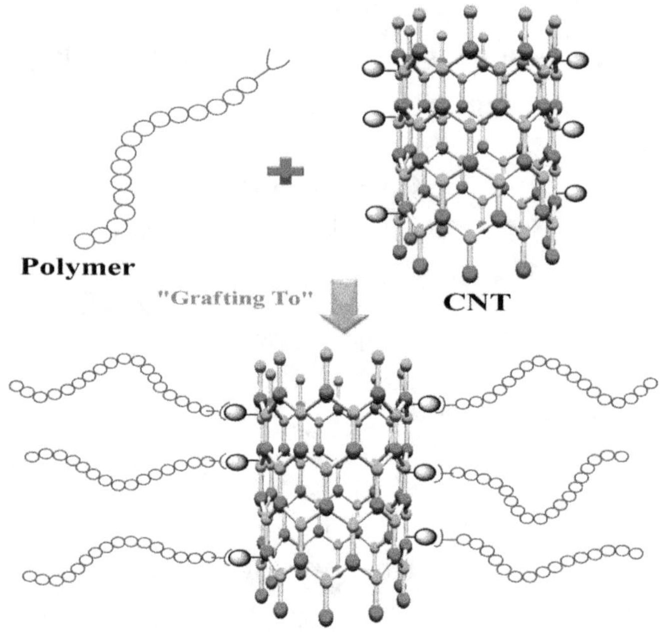

demonstrated a lower molecular weight, they nurture on the nanoparticles and can penetrate aggregated nanoparticles and finally react with the activated sites on the surface of the nanoparticles. The interstitial gap within the nanoparticles becomes moderately incorporated with grafted macromolecular chains. Further, the aggregated nanoparticles become dissociated. The surface of the nanoparticles becomes hydrophobic during addition time. This is significant for the miscibility of the filler and matrix. Figure 8 exhibits the dispersion character of pure and polymer grafted nanoparticles in a polymer matrix (Rong et al., 2001).

PREPARATION OF EPOXY POLYMER NANOCOMPOSITES

Epoxy Polymer Nanocomposites With an Organic Filler

Polymer nanocomposites (PNCs) based on organic nanofiller like carbon nanotube (CNT), graphene oxide (GO) have been reported (Kausar et al., 2016; Abdullah et al., 2015). This section focuses on the employ of CNT and GO as nanofiller in the polymeric system such as epoxy resin. Currently, many researchers had developed high-performance polymer/CNT nanocomposites materials to effectively utilize the

Figure 8. Schematic representation of (a) without polymer grafting, and (b) with polymer grafting

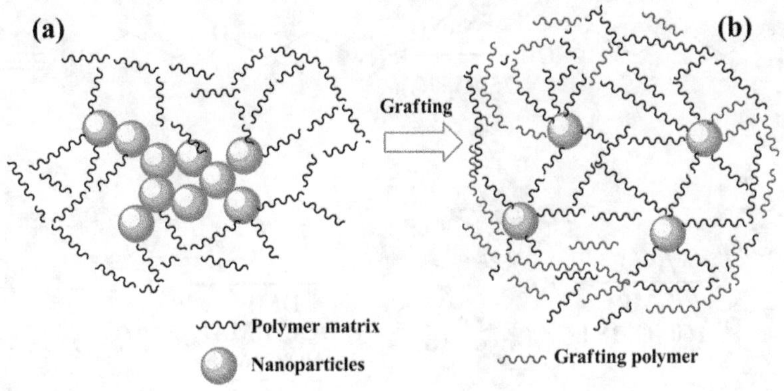

potential of CNTs in various applications like aerospace and automotive (Kausar et al., 2016). The carbon nanoparticles like graphene oxide (GO), single-walled carbon nanotube (SWCNT) and multi-walled carbon nanotubes (MWCNTs) are well-known nanofiller for the high-performance industrial applications.

Carbon Nanotube-Based Epoxy Nanocomposites

The synthesis of epoxy/MWCNTs nanocomposites by the addition of MWCNTs into the epoxy resin as reinforcement has been reported elsewhere (Yeh et al., 2008). The results demonstrated that the mechanical properties like Young's modulus of epoxy matrix enhanced after addition of 5 wt.% MWCNTs contrasted to the virgin epoxy resin. The tensile strength (TS) of the epoxy/MWCNTs nanocomposites was also enhanced by incorporation of 3 wt.% MWCNTs in the epoxy matrix. The incorporation of the higher content of MWCNTs young's modulus increased slowly while the TS reduced owing to the non-uniform dispersion of the epoxy matrix.

Further, the epoxy/MWCNTs nanocomposites were synthesized with the incorporation of the amino-functionalized MWCNTs and the effect on mechanical and thermal properties have been studied. In this investigation, the tensile strength and impact strength both are increased with increasing content of the MWCNTs from 39.1 MPa (virgin epoxy) to 46.0 MPa with the addition of 1.5 wt% oxidative multi-walled carbon nanotube (o-MWCNTs) and 59.2 MPa with the incorporation of 1.5 wt% amine-functionalized multi-walled carbon nanotube(a-MWCNTs). Similarly, the impact strength improved from 5.5kJ/m^2 to 7.2kJ/m^2 and 10.6kJ/m^2 respectively

Figure 9. Synthesis of (a) 2(2-azidoethoxy) ethanol, and (b) MWCNTs-g-CTBN

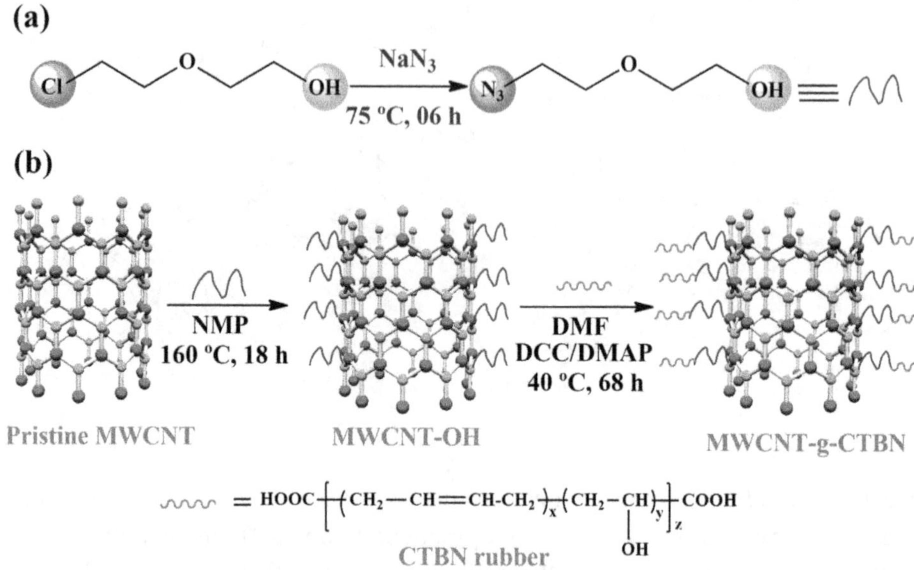

On the other hand, with an increase in the 2 wt% of MWCNTs, the tensile and impact strength is slightly reduced for both systems. The high surface area of CNT effects in a higher viscosity of the epoxy/CNT nanocomposites mixture, when the MWCNTs content was increased. Along with, enhance of viscosity, more voids were generated throughout the synthesis method. This is the new method for the fabrication of epoxy/CNT nanocomposites with exclusive properties for the next generation of nanocomposite materials (Chen et al.; 2008).

In another study, new functionalization of the MWCNTs was prepared by the grafting of carboxyl terminated poly(acrylonitrile-co-butadiene) (MWCNTs-g-CTBN) as shown in Figure 9. The prepared MWCNTs-g-CTBN was incorporated to the diglycidyl ether of bisphenol A epoxy resin to synthesize Epoxy/MWCNTs-g-CTBN nanocomposites.

The TS of the MWCNTs incorporated epoxy nanocomposites enhanced up to 0.2 wt% compared to the virgin epoxy. TS of epoxy nanocomposites decreased beyond 0.2 wt% this is owing to the agglomeration and inhomogeneous dispersion of the MWCNT into the epoxy matrix. On the other hand, the MWCNTs-g-CTBN containing epoxy nanocomposites demonstrated higher tensile and modulus strength than virgin MWCNTs over the entire filler contents as depicted in Table 2. The 3 wt% MWCNTs-g-CTBN epoxy nanocomposites exhibited higher values of tensile strength and modulus of 2.21 ± 0.02GPa and 85.4 ± 3.1 MPa respectively.

Table 2. The tensile and fracture toughness properties of virgin epoxy and its nanocomposites.

Sample	Tensile Strength (MPa)	Tensile Modulus (GPa)	Elongation (%)	K_{IC} (MPa*m$^{1/2}$)
Virgin epoxy	68.4	1.89	5.66	0.74
Epoxy/0.1wt%MWCNT	69.4	1.92	6.01	1.14
Epoxy/0.2wt%MWCNT	76.1	1.98	6.53	1.35
Epoxy/0.3wt%MWCNT	71.6	2.01	6.09	1.15
Epoxy/0.4wt%MWCNT	68.4	1.93	5.75	1.12
Epoxy/0.1wt%MWCNT-g-CTBN	74.1	1.96	6.12	1.39
Epoxy/0.2wt%MWCNT-g-CTBN	79.9	2.05	7.05	147
Epoxy/0.3wt%MWCNT-g-CTBN	85.4	2.21	6.76	1.62
Epoxy/0.4wt%MWCNT-g-CTBN	76.5	2.07	6.96	1.24

(Konnola, et al.)

Table 2 indicated that the fracture toughness, critical stress intensity factor (K_{IC}) of the virgin epoxy, MWCNT and MWCNTs-g-CTBN based nanocomposites. The inclusion of MWCNT and MWCNTs-g-CTBN into the epoxy matrix enhanced the fracture toughness. The virgin epoxy exhibited the K_{IC} value of 0.74 MPa.m$^{1/2}$ which is raised up to 1.35 MPa m$^{1/2}$ via incorporating 0.2 wt% MWCNTs. Similarly, the MWCNTs-g-CTBN containing epoxy nanocomposites enhanced the K_{IC} value extensively at 117% loading of 0.3 wt% filler. On the other hand, increasing the concentration of MWCNT and MWCNTs-g-CTBN in the epoxy matrix decreased the K_{IC} value owing to the existence of agglomeration and defects. The dynamic mechanical analysis (DMA) and thermal gravimetric analysis also confirmed that the MWCNT and MWCNTs-g-CTBN modified epoxy nanocomposites revealed higher storage modulus, glass transition temperature and thermal properties compared to virgin epoxy. Finally, the MWCNTs-g-CTBN is a useful reinforcement for epoxy systems for a high-performance application (Konnola, et al.).

Graphene Oxide (GO) Based Epoxy Nanocomposites

The constituent of the GO is based on carbon with excellent potential and low cost. The epoxy/GO nanocomposite contains higher tensile strength and modulus. They can be used in various mechanical, civil and aerospace applications. In epoxy

Table 3. The mechanical properties of epoxy and its nanocomposites (Abdullah et al. 2014)

Epoxy/GO Nanocomposites (Vol.%)	Tensile Strength (MPa)	Young's Modulus (MPa)	Elongation (%)	Impact Strength (J/mm^2)
Virgin epoxy	7	115	38	2.56
Epoxy/GO (68.9/1.5 vol.%)	13	167	55	3.2
Epoxy/GO (67.9/3.0 vol.%)	9	178	56	17.3
Epoxy/GO (66.8/4.5 vol.%)	8	192	58	39.7
Epoxy/GO (65.8/6.0 vol.%)	7.5	206	65	32.0

nanocomposites systems, GO is employed as a filler for increasing the mechanical performance of polymer matrix (Kausar et al., 2017; (Abdullah et al. 2015).

Fabrication of epoxy/GO nanocomposites for a various percentage of GO (1.5-6 vol.%) was added into the epoxy matrix and studied the effect of GO contents (Abdullaha et al. 2014). As a result, after the addition of GO, the mechanical properties enhanced like tensile strength, and impact of epoxy/GO nanocomposites as summarized in Table 3. Table 3 shows that the incorporation of GO (1.5 vol.%) to the epoxy matrix enhanced the TS from 7-13 MPa. Further, the increase in GO content up to 6.0 vol.% reduces the TS to 7.5 MPa. The Youngs modulus of the epoxy nanocomposites constantly enhanced from 115-206 MPa after incorporation of 0 vol.% to 6 vol.% of GO. This is ascribed to the improved interfacial adhesion and compatibility amongst GO and epoxy matrix. Table 3 demonstrated the effect of GO content in the elongation (%). The increased GO content surprisingly enhanced the elongation of 38%.On the other hand, impact strength enhanced from 2.56 J/mm^2 to 39.7 J/mm^2 at 4.5 vol.% GO content as shown in Table 3 owing to the formation of chemical bonding between the epoxy resin and GO filler. Further, the impact strength reduced to 32 J/mm^2 at 6 vol.% GO content due to the inhomogeneous dispersion. Finally, the epoxy/GO nanocomposites exhibited potential application in the aerospace and automotive industries.

The synthesis of epoxy nanocomposites with the incorporation of GO and diglycidyl ether of bisphenol A functionalized GO (DGEBA/F/GO) with various percentage of filler loading was shown in Figure 10 (Wan et al., 2014). The DGEBA layer successfully enhanced the compatibility and dispersion of GO in epoxy matrix specimens. As a result, the tensile strength confirmed that the DGEBA/F/GO/epoxy nanocomposites demonstrated higher tensile strength and modulus compared to the virgin epoxy and epoxy/GO nanocomposites as depicted in Table 4.

Figure 10. Synthetic routes of the epoxy matrix and DGEBA–f–GO nanocomposites

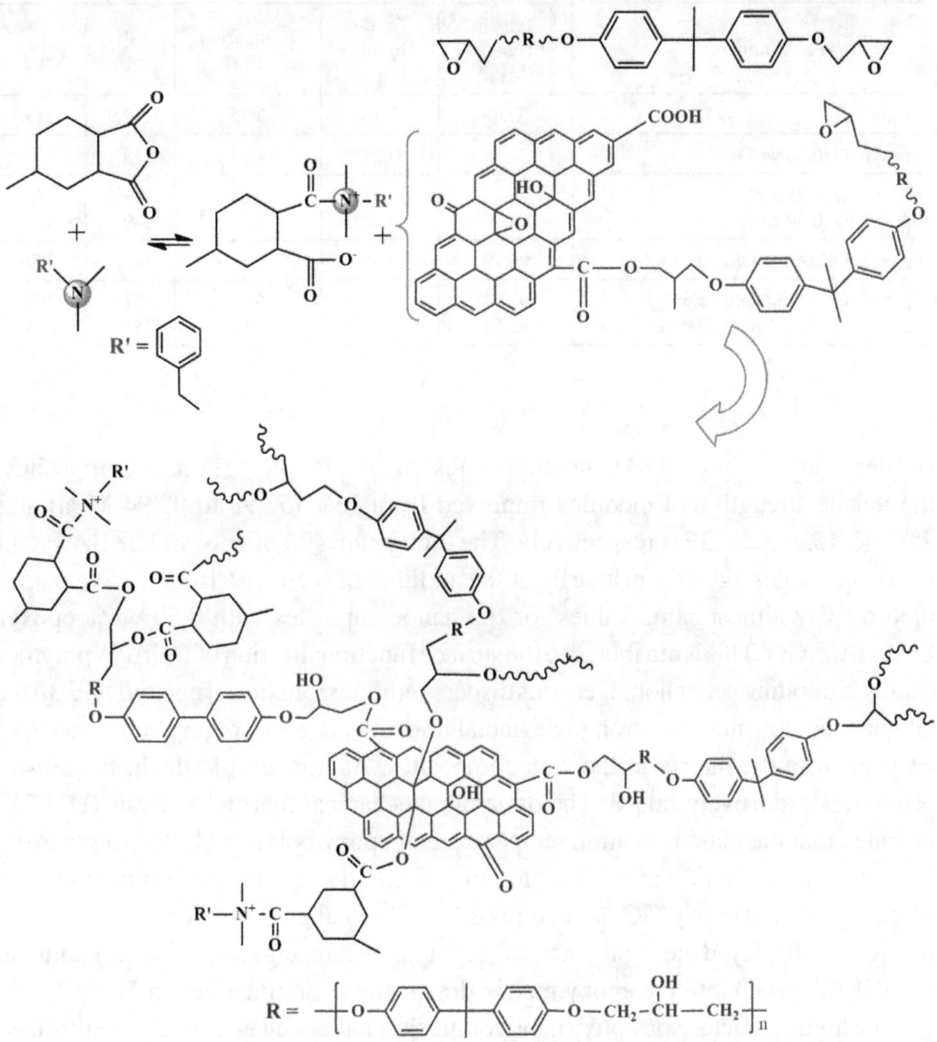

The TS enhanced is 35% after addition of 0.10 wt.% GO, and slightly decreased due to the increase of GO content in epoxy/GO nanocomposites. On the other hand, tensile modulus also increased (7%) up to loading 0.50 wt.% GO. Elongation % slightly enhanced with the inclusion of low wt.% of 0.10 GO and further reduces with the increasing of GO content as summarized in Table 4.

Conversely, the TS and modulus of DGEBA/F/GO nanocomposites enhance with increasing loading of filler content from 0-0.50 wt% as shown in Table 4. However, the tensile property of the epoxy/DGEBA/F/GO nanocomposites is approximately

Table 4. The mechanical and thermal properties of virgin epoxy and its nanocomposites.

Sample	Tensile Strength (MPa)	Tensile Modulus (GPa)	Elongation (%)	T_g (°C)	T_d (°C)
Virgin epoxy	52.98	3.15	3.75	147.9	343
Epoxy/GO (0.10 wt%)	71.54	3.27	4.23	148.1	339
Epoxy/GO (0.25 wt%)	68.02	3.32	3.85	149.5	358
Epoxy/GO (0.50 wt%)	64.79	3.36	3.51	149.5	355
Epoxy/DGEBA/F/GO(0.10 wt%)	94.79	3.35	6.45	148.9	359
Epoxy/DGEBA/F/GO(0.25 wt%)	92.94	3.56	6.32	151.0	363
Epoxy/DGEBA/F/GO(0.50 wt%)	85.51	3.67	5.86	151.8	373

(Wan et al., 2014)

similar to that of epoxy/GO nanocomposites. In the case of epoxy nanocomposites, the tensile strength and modulus improved from 13% (52.98 to 92.94 MPa) and 75%, (3.15 to 3.56 GPa) respectively. The elongation (%) of epoxy/DGEBA/F/GO nanocomposites enhance primarily at the addition of 0.10 wt% from 3.75 to 6.45% then display almost same values for the nanocomposites with 0.50 wt.% epoxy/DGEBA/F/GO. This is attributed to the surface functionalization of DGEBA polymer chains, exhibiting better homogeneous dispersion and exfoliation of epoxy/DGEBA/F/GO nanocomposites and strong interfacial bonding are encouraging enhanced load transfer from the matrix to the nanocomposites. As a result of which, the tensile properties extensively raised. The dynamic mechanical thermal analysis (DMTA) revealed that the glass transition temperature of epoxy polymer shifted towards the higher temperature (T_g) after inclusion of DGEBA/F/GO content compared to the virgin epoxy and epoxy/GO nanocomposites. Thermal gravimetric analysis (TGA) results also displayed the higher thermal degradation temperature (T_d) in the inclusion of DGEBA/F/GO into the epoxy matrix drastically as summarized in Table 4.

The high characteristic epoxy nanocomposites have been effectively synthesized by the addition of liquid crystal grafted epoxy resin graphene oxide (LCE/g/GO) (Lu et al., 2013). The properties of Epoxy/LCE/g/GO nanocomposites were based on the interfacial interaction amongst LCE/g/GO and epoxy matrix. The results demonstrated that the incorporation of LCE/g/GO extensively enhanced its mechanical and thermal properties. Table 5 shows that the flexural strength and modulus of the epoxy nanocomposites improved with enhancing of LCE/g/GO content. The highest value of flexural strength and modulus obtained 135 MPa and 2916 MPa with the inclusion of 5 wt% LCE/g/GO into the epoxy matrix. Similarly, the impact strength also achieved higher value 35.06 kJm^{-2} it is two times higher than that of virgin epoxy of 17.49 kJm^{-2} with the incorporation of 5 wt% LCE/g/GO content while the

Table 5. Mechanical properties of EP/(LCE/g/GO) nanocomposites.

Sample	Flexural Strength (MPa)	Flexural Modulus (MPa)	Impact Strength (kJ M^{-2})
Virgin epoxy/0wt%	79.35	2140	17.49
Epoxy/(LCE/g/GO)1wt%	110.45	2524	32.41
Epoxy/(LCE/g/GO)3wt%	126.61	2748	33.82
Epoxy/(LCE/g/GO)5wt%	135.00	2916	35.06
Epoxy/(LCE/g/GO)7wt%	103.92	2714	30.43

(Lu et al., 2013)

addition of 7 wt% the impact strength slightly decreased but it is higher than virgin epoxy. Furthermore, the LCE/g/GO nanosheet exhibit remarkable potential as new and valuable additives for enhancing the mechanical and thermal properties in PNCs.

EPOXY POLYMER NANOCOMPOSITES WITH INORGANIC NANOFILLER

Nanoclay Nanoparticles

The epoxy/clay nanocomposites with improved tensile and fracture toughness without altering its strength and modulus, by adding modified copolymer to lower the interaction of clay with the epoxy curing route (Wang et al 2015), leading to an interpenetrating network (IPN) structure inside epoxy system, resulting in to a double IPN structure as shown in Figure 11.

The epoxy matrix becomes even more strong and tough due to these interlocked double IPN structures. At low content of modified clay, improved the various properties such as T_g, TS, modulus, elongation (%), fracture toughness (K_{IC}), storage modulus, and thermal stability enhances to altering degrees. At 0.5 wt%, pure clay and epoxy nanocomposite reveals 30% and 25% decrease in highest compressive extension and critical fracture load respectively, in contrast to pure epoxy, signifying that nanocomposite became more brittle compared to virgin epoxy after incorporation of the pure clay. In the absence of proper compatibilizers, the filler and the matrix shows weak interfacial interaction. However, in the existence of modified clay, epoxy nanocomposites show the maximum compressive strain, but the critical fracture load initially improves with clay amount and then lowers at 1.5 wt% of the modified clay. At 1 wt% clay loading, tensile strength and modulus, elongation (%) plus K_{1C} value of the epoxy/clay nanocomposite attain 38%, 5%, 64%, 93%, respectively, indicating

Figure 11. Schematic depiction of epoxy/Clay nanocomposites

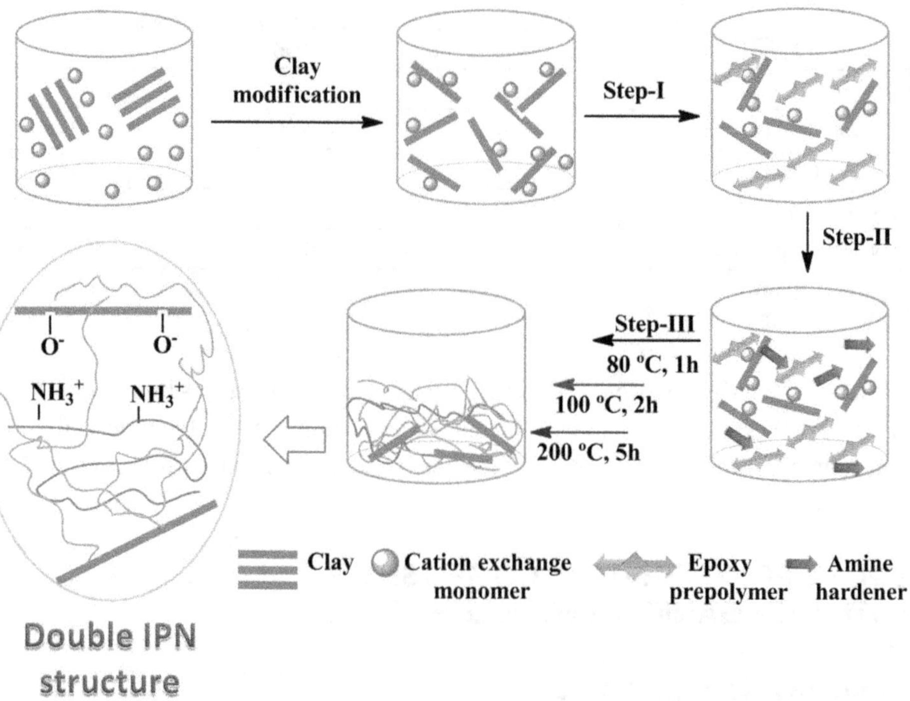

a significant increment of tensile and fracture toughness whereas retaining the normal strength. The improve in tensile modulus upon the inclusion of nanofillers is frequently supposed, but the existence of the rigid filler generally decreases the ductility of the flexible polymer matrix, foremost to declined elongation (%) and K_{IC} value (Piscitelli et al., 2012).

Epoxy nanocomposites can also be synthesised by a novel liquid-liquid method which enhanced the dispersion and exfoliation of the organoclay in the epoxy matrix (Wang et al. 2006). In contrast to nanocomposites prepared via the normal direct-mixing method, this method exhibits an immense increase in viscosity. In contrast to pure epoxy–amine cured system, the 5 wt % organoclay nanocomposite cured samples revealed a 45% enhance in highest strength, 10% rise in yield strength, and a modulus increased to 26%.

Titanium Dioxide (TiO_2) Nanoparticles

Epoxy resins incorporated with titanium dioxide (TiO_2) particles exhibit variant properties owing to various reasons like processing methods, physical and chemical character of TiO_2 nanoparticles. Though, in comparison to the virgin epoxy polymer, the enhancement in mechanical properties and glass transition temperature (T_g) has been found for the TiO_2 incorporated epoxy nanocomposites (Na et al., 1999; Werzel et al., 2006; Zhuo et al., 2010; Al-Turaif et al., 2010 Polizos et al., 2011)

The epoxy/TiO_2 nanocomposite was prepared, by adding the micron and nano-sized TiO_2 particles (Singh et al., 2011). Results show that the compared to virgin epoxy, TiO_2 nanofillers incorporated epoxy resin exhibits improved tensile strength and fracture properties revealing the fact that TiO_2 nanofillers in the epoxy matrix has significantly enhanced the mechanical properties of the epoxy nanocomposites. At 4 wt % TiO_2, an optimum increase of 32.57% and 44.95% in tensile strength and fracture toughness, respectively was observed for micron-sized TiO_2 reinforced epoxy nanocomposites. On the other hand, for nano-sized TiO_2 reinforced epoxy nanocomposites, the optimum increase of up to 43.7% and 141.85% respectively for tensile strength and fracture toughness was observed at 6 wt % TiO_2. Whereas, for 4 wt% micron and nano-sized TiO_2 respectively, in the glassy region, the storage modulus of filled epoxy composites exhibits maximum improvement of around 4.73% and 32.88% respectively in comparison to neat epoxy at 40°C. Around individual particles, improved connectivity of yielded microzones leads to a maximum reinforcing effect created via the TiO_2 nanoparticles as contrasted to that exhibited via usual particles or nanoparticles clusters in the microns scale (Ahmed et al., 1990). The variation in the morphological structure improves the interaction amongst the nanoparticle and the epoxy matrix at the particle-matrix interface, which also resulted in improved toughness (Kinloch et al., 2005). But, more rise of TiO_2 content in the epoxy matrix leads to higher cluster size finally causing cracking, which decreases the toughness of the nanocomposite.

The un-functionalized TiO_2 nanoparticles incorporated epoxy resin was synthesised by optimized ultrasonic-assisted dispersion method as depicted in Figure 12 (Goyat et al.; 2017). The results reveal that there is an interlink between nanoparticles loading ratio, inter-particle spacing and agglomerate size of the particles on the tensile properties and glass transition temperature (T_g) of the developed nanocomposites. Considerable enhancement in tensile strength (TS) and modulus without altering the toughness was seen for the developed nanocomposites, in comparison to virgin epoxy resins, for 10 wt% particles, around 26% and 18% enhancement in TS and strain-to-break %, correspondingly, was detected. However, for 5 wt% particles, the highest enhancement of around 54% for tensile toughness was found. Results indicate that a better particle-matrix interaction leads to the improvement of the mechanical properties owing to various toughening mechanisms like particle pull out, crack deflection, and plastic deformation. The strength decreases at highest nanoparticles loading ratio for example 15 to 20 wt% because of the agglomeration of nanoparticles. However, the particle-particle interaction plays a vital role instead of particle-matrix polymer interaction.

Figure 12. Schematic representation preparation of epoxy/TiO_2 nanocomposites

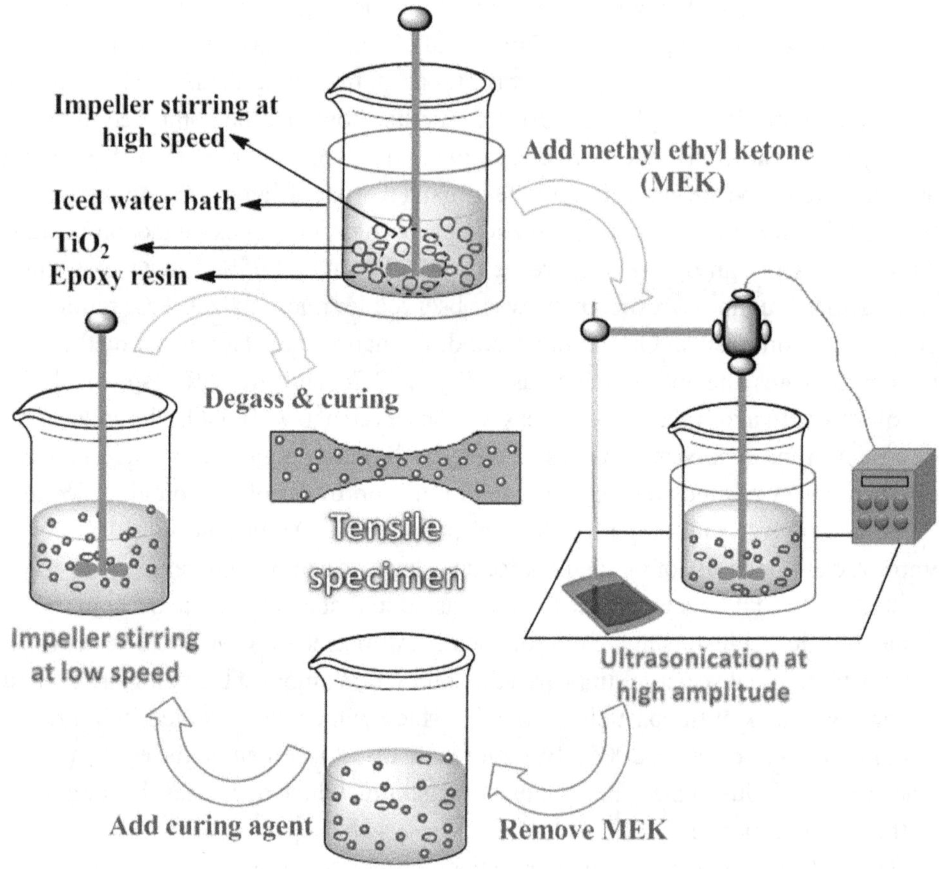

AEROSPACE AND AUTOMOTIVE APPLICATION OF EPOXY NANOCOMPOSITES

- **Aerospace:** Generally, the materials for aerospace application are subject to various climate conditions as well as higher humidity, large temperature variations, and several types of mechanical stresses like tension, torsion creep, and compression. However, traditional materials like steel, aluminium, and titanium can achieve a few of the demands, but they cannot compromise a lower weight (Guadagno et al., 2014). Recently, the requirement of the innovative polymer nanocomposites materials increased in the aerospace industry extensively as the load moving components of latest aircraft like Airbus 350, F-35, and Boeing 787 needs efficient weight reduction (Sugita et al., 2010).

Figure 13. Epoxy nanocomposites application in aerospace and automotive areas

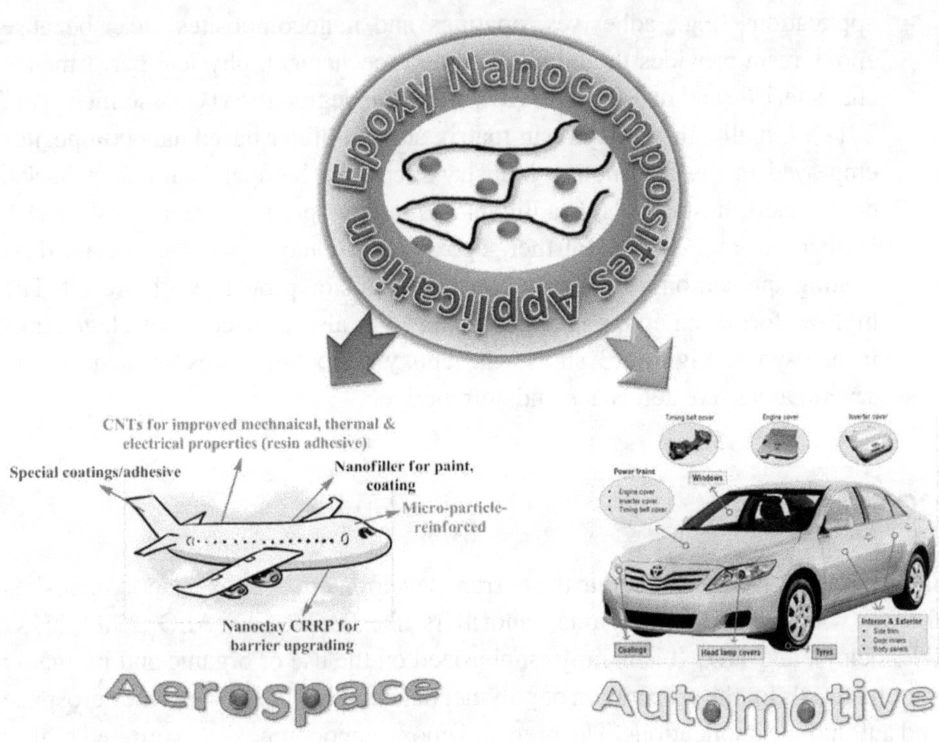

Thermosetting epoxy-based PNCs are one of the most generally employed aeronautic materials in the aerospace application, owing to the tremendous mechanical properties, chemical and electrical resistance and low shrinkage on curing (Toldy et al., 2011). Incorporation of multifunctionality has become a very important characteristic in the aerospace industry in current years. The multifunctional epoxy nanocomposites increased the mechanical and thermal performance according to the aerospace sector demand (Paipetis et al., 2012, Kumar et al., 2019). Consequently, the utilization of the latest innovative epoxy nanocomposites with additional functional groups without negotiation of structural properties is needed. In the last few years, innovative nanofiller materials have emerged as tremendous contenders to bestow final products with increased structural properties plus multifunctionality. Because nanofiller demonstrated higher TS with a high aspect ratio and outstanding electrical conductivity (e.g., CNTs), which make them potential materials for the influence of a new invention of nano-reinforced composites structures in aerospace applications to substitute the traditional materials.

- **Automotive:** Epoxy resin is widely utilized thermosetting resin for automotive applications (e.g., adhesives, coatings and nanocomposites etc.,) because epoxy resin provides the higher T_g, good mechanical, physical performance characteristic and resistance to environmental degradation (Ghassemieh et al., 2011). Usually, the epoxy resin matrix and nanofiller based nanocomposites employed in the automotive industry such as a bumper beam, seat backs, door panels, dashboards, headliners and interior parts (Davoodi et al., 2010; Holbery et al., 2006)). Further, epoxy resin nanocomposites are used as coating application for enhanced the anticorrosion properties of aircraft. The high-performance epoxy nanocomposites are also utilized in fuselage sinks in aerospace. Figure 13 shows the epoxy nanocomposites applications in several areas like aerospace and automotive.

CONCLUSION

In this chapter, we can deliberate the current development of epoxy nanocomposites for use with organic and organic nanofillers like CNTs, GO, $Al_2O_3.2SiO_2.2H_2O$ (Nanoclay), and $TiO_{2.}$ It is mainly emphasized on the use of organic and inorganic nanomaterials for the preparation of polymer nanocomposites towards the aerospace and automobile applications. The prepared epoxy nanocomposites exhibited higher mechanical and thermal properties with the addition of nanofillers to enhance the dispersion properties of nanofillers and improved the interfacial interaction among nanofillers and epoxy matrix. The innovative epoxy nanocomposites demonstrated higher potential to explore new industrial applications.

REFERENCES

Abdullah, S. I., & Ansari, M. N. M. (2015). Mechanical properties of graphene oxide (GO)/epoxy composites. *Hbrc Journal*, *11*(2), 151–156. doi:10.1016/j.hbrcj.2014.06.001

Acebo, C., Fernández-Francos, X., Messori, M., Ramis, X., & Serra, À. (2014). Novel epoxy-silica hybrid coatings by using ethoxysilyl-modified hyperbranched poly (ethyleneimine) with improved scratch resistance. *Polymer*, *55*(20), 5028–5035. doi:10.1016/j.polymer.2014.08.021

Ahmed, S., & Jones, F. R. (1990). A review of particulate reinforcement theories for polymer composites. *Journal of Materials Science*, *25*(12), 4933–4942. doi:10.1007/BF00580110

Al-Turaif, H. A. (2010). Effect of nano TiO2 particle size on mechanical properties of cured epoxy resin. *Progress in Organic Coatings*, *69*(3), 241–246. doi:10.1016/j.porgcoat.2010.05.011

Asmatulu, R., Khan, W. S., Reddy, R. J., & Ceylan, M. (2015). Synthesis and analysis of injection-molded nanocomposites of recycled high-density polyethylene incorporated with graphene nanoflakes. *Polymer Composites*, *36*(9), 1565–1573. doi:10.1002/pc.23063

Bose, S., Kuila, T., Uddin, M. E., Kim, N. H., Lau, A. K., & Lee, J. H. (2010). In-situ synthesis and characterization of electrically conductive polypyrrole/graphene nanocomposites. *Polymer*, *51*(25), 5921–5928. doi:10.1016/j.polymer.2010.10.014

Chan, C. M., Wu, J., Li, J. X., & Cheung, Y. K. (2002). Polypropylene/calcium carbonate nanocomposites. *Polymer*, *43*(10), 2981–2992. doi:10.1016/S0032-3861(02)00120-9

Chandrasekaran, S., Seidel, C., & Schulte, K. (2013). Preparation and characterization of graphite nano-platelet (GNP)/epoxy nano-composite: Mechanical, electrical and thermal properties. *European Polymer Journal*, *49*(12), 3878–3888. doi:10.1016/j.eurpolymj.2013.10.008

Chatterjee, A., & Islam, M. S. (2008). Fabrication and characterization of TiO2–epoxy nanocomposite. *Materials Science and Engineering A*, *487*(1-2), 574–585. doi:10.1016/j.msea.2007.11.052

Chee, W. K., Lim, H. N., Huang, N. M., & Harrison, I. (2015). Nanocomposites of graphene/polymers: A review. *RSC Advances*, *5*(83), 68014–68051. doi:10.1039/C5RA07989F

Chen, X., Wang, J., Lin, M., Zhong, W., Feng, T., Chen, X., & Xue, F. (2008). Mechanical and thermal properties of epoxy nanocomposites reinforced with amino-functionalized multi-walled carbon nanotubes. *Materials Science and Engineering A*, *492*(1-2), 236–242. doi:10.1016/j.msea.2008.04.044

Choi, K. S., Liu, F., Choi, J. S., & Seo, T. S. (2010). Fabrication of free-standing multilayered graphene and poly (3,4-ethylenedioxythiophene) composite films with enhanced conductive and mechanical properties. *Langmuir*, *26*(15), 12902–12908. doi:10.1021/la101698j PMID:20617852

Cui, L. J., Geng, H. Z., Wang, W. Y., Chen, L. T., & Gao, J. (2013). Functionalization of multi-wall carbon nanotubes to reduce the coefficient of the friction and improve the wear resistance of multi-wall carbon nanotube/epoxy composites. *Carbon*, *54*, 277–282. doi:10.1016/j.carbon.2012.11.039

Davoodi, M. M., Sapuan, S. M., Ahmad, D., Ali, A., Khalina, A., & Jonoobi, M. (2010). Mechanical properties of hybrid kenaf/glass reinforced epoxy composite for passenger car bumper beam. *Materials & Design, 31*(10), 4927–4932. doi:10.1016/j.matdes.2010.05.021

Dennis, H., Hunter, D. L., Chang, D., Kim, S., White, J. L., Cho, J. W., & Paul, D. R. (2001). Effect of melt processing conditions on the extent of exfoliation in organoclay-based nanocomposites. *Polymer, 42*(23), 9513–9522. doi:10.1016/S0032-3861(01)00473-6

Dreyer, D. R., Park, S., Bielawski, C. W., & Ruoff, R. S. (2010). The chemistry of graphene oxide. *Chemical Society Reviews, 39*(1), 228–240. doi:10.1039/B917103G PMID:20023850

Ge, T., Kalathi, J. T., Halverson, J. D., Grest, G. S., & Rubinstein, M. (2017). Nanoparticle motion in entangled melts of linear and nonconcatenated ring polymers. *Macromolecules, 50*(4), 1749–1754. doi:10.1021/acs.macromol.6b02632 PMID:28392603

Ghassemieh, E. (2011). Materials in automotive application, state of the art and prospects. In New trends and developments in automotive industry. InTechOpen. doi:10.5772/13286

Goyat, M. S., Rana, S., Halder, S., & Ghosh, P. K. (2018). Facile fabrication of epoxy-TiO2 nanocomposites: A critical analysis of TiO2 impact on mechanical properties and toughening mechanisms. *Ultrasonics Sonochemistry, 40*, 861–873. doi:10.1016/j.ultsonch.2017.07.040 PMID:28946497

Gu, H., Guo, J., He, Q., Tadakamalla, S., Zhang, X., Yan, X., & Guo, Z. (2013). Flame-retardant epoxy resin nanocomposites reinforced with polyaniline-stabilized silica nanoparticles. *Industrial & Engineering Chemistry Research, 52*(23), 7718–7728. doi:10.1021/ie400275n

Guadagno, L., Raimondo, M., Vittoria, V., Vertuccio, L., Naddeo, C., Russo, S., & Tucci, V. (2014). Development of epoxy mixtures for application in aeronautics and aerospace. *RSC Advances, 4*(30), 15474–15488. doi:10.1039/C3RA48031C

Guo, Z., Henry, L. L., Palshin, V., & Podlaha, E. J. (2006). Synthesis of poly (methyl methacrylate) stabilized colloidal zero-valence metallic nanoparticles. *Journal of Materials Chemistry, 16*(18), 1772–1777. doi:10.1039/B515565G

Guo, Z., Wei, S., Shedd, B., Scaffaro, R., Pereira, T., & Hahn, H. T. (2007). Particle surface engineering effect on the mechanical, optical and photoluminescent properties of ZnO/vinyl-ester resin nanocomposites. *Journal of Materials Chemistry*, *17*(8), 806–813. doi:10.1039/B613286C

He, F., Fan, J., Ma, D., Zhang, L., Leung, C., & Chan, H. L. (2010). The attachment of Fe3O4 nanoparticles to graphene oxide by covalent bonding. *Carbon*, *48*(11), 3139–3144. doi:10.1016/j.carbon.2010.04.052

Holbery, J., & Houston, D. (2006). Natural-fiber-reinforced polymer composites in automotive applications. *JOM*, *58*(11), 80–86. doi:10.100711837-006-0234-2

Hsiao, K. T., Alms, J., & Advani, S. G. (2003). Use of epoxy/multiwalled carbon nanotubes as adhesives to join graphite fibre reinforced polymer composites. *Nanotechnology*, *14*(7), 791–793. doi:10.1088/0957-4484/14/7/316

Huang, W., Wang, S., Guo, C., Yang, X., Li, Y., & Tu, Y. (2014). Synthesis and characterization of well-defined poly (l-lactide) functionalized graphene oxide sheets with high grafting ratio prepared through click chemistry and supramolecular interactions. *Polymer*, *55*(18), 4619–4626. doi:10.1016/j.polymer.2014.07.014

Jannapu Reddy, R. (2010). *Preparation, characterization and properties of injection molded graphene nanocomposites* (Doctoral dissertation). Wichita State University.

Jyotishkumar, P., Koetz, J., Tiersch, B., Strehmel, V., Özdilek, C., Moldenaers, P., & Thomas, S. (2009). Complex phase separation in poly (acrylonitrile-butadiene-styrene)-modified epoxy/4, 4′-diaminodiphenyl sulfone blends: Generation of new micro-and nanosubstructures. *The Journal of Physical Chemistry B*, *113*(16), 5418–5430. doi:10.1021/jp8094566 PMID:19331324

Kango, S., Kalia, S., Celli, A., Njuguna, J., Habibi, Y., & Kumar, R. (2013). Surface modification of inorganic nanoparticles for development of organic–inorganic nanocomposites—A review. *Progress in Polymer Science*, *38*(8), 1232–1261. doi:10.1016/j.progpolymsci.2013.02.003

Kao, J., Thorkelsson, K., Bai, P., Rancatore, B. J., & Xu, T. (2013). Toward functional nanocomposites: Taking the best of nanoparticles, polymers, and small molecules. *Chemical Society Reviews*, *42*(7), 2654–2678. doi:10.1039/C2CS35375J PMID:23192158

Kausar, A., Rafique, I., Anwar, Z., & Muhammad, B. (2016). Perspectives of epoxy/graphene oxide composite: Significant features and technical applications. *Polymer-Plastics Technology and Engineering*, *55*(7), 704–722. doi:10.1080/03602559.2015.1098700

Kausar, A., Rafique, I., & Muhammad, B. (2017). Aerospace application of polymer nanocomposite with carbon nanotube, graphite, graphene oxide, and nanoclay. *Polymer-Plastics Technology and Engineering*, *56*(13), 1438–1456. doi:10.1080/03602559.2016.1276594

Khan, W. S., Hamadneh, N. N., & Khan, W. A. (2016). *Polymer nanocomposites–synthesis techniques, classification and properties. Science and applications of Tailored Nanostructures: One Central Press*. OCP.

Kinloch, A. J., Mohammed, R. D., Taylor, A. C., Eger, C., Sprenger, S., & Egan, D. (2005). The effect of silica nano particles and rubber particles on the toughness of multiphase thermosetting epoxy polymers. *Journal of Materials Science*, *40*(18), 5083–5086. doi:10.100710853-005-1716-2

Konnola, R., & Joseph, K. (2016). Effect of side-wall functionalisation of multi-walled carbon nanotubes on the thermo-mechanical properties of epoxy composites. *RSC Advances*, *6*(28), 23887–23899. doi:10.1039/C6RA00080K

Kumar, K., Ghosh, P. K., & Kumar, A. (2016). Improving mechanical and thermal properties of TiO2-epoxy nanocomposite. *Composites. Part B, Engineering*, *97*, 353–360. doi:10.1016/j.compositesb.2016.04.080

Kumar, S., Krishnan, S., Samal, S. K., Mohanty, S., & Nayak, S. K. (2019). Polymer Nanocomposites Coating for Anticorrosion Application. In N. Ramdani (Ed.), *Polymer Nanocomposites for Advanced Engineering and Military Applications* (pp. 254–294). Hershey, PA: IGI Global; doi:10.4018/978-1-5225-7838-3.ch009

Kumar, S., Samal, S. K., Mohanty, S., & Nayak, S. K. (2017). Synthesis and Characterization of Nanoclay-Reinforced Trifunctional "Bioresin-Modified" Epoxy Blends Enhanced with Mechanical and Thermal Properties. *ChemistrySelect*, *2*(35), 11445–11455. doi:10.1002lct.201702041

Lu, S., Li, S., Yu, J., Yuan, Z., & Qi, B. (2013). Epoxy nanocomposites filled with thermotropic liquid crystalline epoxy grafted graphene oxide. *RSC Advances*, *3*(23), 8915–8923. doi:10.1039/c3ra40404h

Mittal, V. (2014). Functional polymer nanocomposites with graphene: A review. *Macromolecular Materials and Engineering*, *299*(8), 906–931. doi:10.1002/mame.201300394

Ng, C. B., Schadler, L. S., & Siegel, R. W. (1999). Synthesis and mechanical properties of TiO2-epoxy nanocomposites. *Nanostructured Materials*, *12*(1-4), 507–510. doi:10.1016/S0965-9773(99)00170-1

Paipetis, A., & Kostopoulos, V. (Eds.). (2012). *Carbon nanotube enhanced aerospace composite materials: a new generation of multifunctional hybrid structural composites* (Vol. 188). Springer Science & Business Media.

Pham, T. A., Kim, J. S., Kim, D., & Jeong, Y. T. (2012). Facile preparation of water-dispersible graphene nanosheets by covalent functionalization with poly (3-aminobenzene sulfonic acid). *Polymer Engineering and Science*, *52*(9), 1854–1861. doi:10.1002/pen.23136

Piscitelli, F., Scamardella, A. M., Romeo, V., Lavorgna, M., Barra, G., & Amendola, E. (2012). Epoxy composites based on amino-silylated MMT: The role of interfaces and clay morphology. *Journal of Applied Polymer Science*, *124*(1), 616–628. doi:10.1002/app.35015

Polizos, G., Tuncer, E., Sauers, I., & More, K. L. (2011). Physical properties of epoxy resin/titanium dioxide nanocomposites. *Polymer Engineering and Science*, *51*(1), 87–93. doi:10.1002/pen.21783

Potts, J. R., Dreyer, D. R., Bielawski, C. W., & Ruoff, R. S. (2011). Graphene-based polymer nanocomposites. *Polymer*, *52*(1), 5–25. doi:10.1016/j.polymer.2010.11.042

Qi, B., Zhang, Q. X., Bannister, M., & Mai, Y. W. (2006). Investigation of the mechanical properties of DGEBA-based epoxy resin with nanoclay additives. *Composite Structures*, *75*(1-4), 514–519. doi:10.1016/j.compstruct.2006.04.032

Qing, Y., Wang, X., Zhou, Y., Huang, Z., Luo, F., & Zhou, W. (2014). Enhanced microwave absorption of multi-walled carbon nanotubes/epoxy composites incorporated with ceramic particles. *Composites Science and Technology*, *102*, 161–168. doi:10.1016/j.compscitech.2014.08.006

Rodríguez-González, C., Martínez-Hernández, A. L., Castaño, V. M., Kharissova, O. V., Ruoff, R. S., & Velasco-Santos, C. (2012). Polysaccharide nanocomposites reinforced with graphene oxide and keratin-grafted graphene oxide. *Industrial & Engineering Chemistry Research*, *51*(9), 3619–3629. doi:10.1021/ie200742x

Rong, M. Z., Zhang, M. Q., Zheng, Y. X., Zeng, H. M., Walter, R., & Friedrich, K. (2001). Structure–property relationships of irradiation grafted nano-inorganic particle filled polypropylene composites. *Polymer*, *42*(1), 167–183. doi:10.1016/S0032-3861(00)00325-6

Shi, X., Nguyen, T. A., Suo, Z., Liu, Y., & Avci, R. (2009). Effect of nanoparticles on the anticorrosion and mechanical properties of epoxy coating. *Surface and Coatings Technology*, *204*(3), 237–245. doi:10.1016/j.surfcoat.2009.06.048

Singh, N. B., Rai, S., & Agarwal, S. (2014). Polymer nanocomposites and Cr (VI) removal from water. *Nanosci. Technol*, *1*, 1–10.

Singh, S. K., Singh, S., Kumar, A., & Jain, A. (2017). Thermo-mechanical behavior of TiO2 dispersed epoxy composites. *Engineering Fracture Mechanics*, *184*, 241–248. doi:10.1016/j.engfracmech.2017.09.005

Sugita, Y., Winkelmann, C., & La Saponara, V. (2010). Environmental and chemical degradation of carbon/epoxy lap joints for aerospace applications, and effects on their mechanical performance. *Composites Science and Technology*, *70*(5), 829–839. doi:10.1016/j.compscitech.2010.01.021

Tantis, I., Psarras, G. C., & Tasis, D. (2012). Functionalized graphene--poly (vinyl alcohol) nanocomposites: Physical and dielectric properties. *Express Polymer Letters*, *6*(4), 283–292. doi:10.3144/expresspolymlett.2012.31

Tao, P., Viswanath, A., Li, Y., Siegel, R. W., Benicewicz, B. C., & Schadler, L. S. (2013). Bulk transparent epoxy nanocomposites filled with poly (glycidyl methacrylate) brush-grafted TiO2 nanoparticles. *Polymer*, *54*(6), 1639–1646. doi:10.1016/j.polymer.2013.01.032

Thangavel, E., Ramasundaram, S., Pitchaimuthu, S., Hong, S. W., Lee, S. Y., Yoo, S. S., ... Kang, Y. S. (2014). Structural and tribological characteristics of poly (vinylidene fluoride)/functionalized graphene oxide nanocomposite thin films. *Composites Science and Technology*, *90*, 187–192. doi:10.1016/j.compscitech.2013.11.007

Tian, W., Liu, L., Meng, F., Liu, Y., Li, Y., & Wang, F. (2014). The failure behaviour of an epoxy glass flake coating/steel system under marine alternating hydrostatic pressure. *Corrosion Science*, *86*, 81–92. doi:10.1016/j.corsci.2014.04.038

Tiwari, J. N., Tiwari, R. N., & Kim, K. S. (2012). Zero-dimensional, one-dimensional, two-dimensional and three-dimensional nanostructured materials for advanced electrochemical energy devices. *Progress in Materials Science*, *57*(4), 724–803. doi:10.1016/j.pmatsci.2011.08.003

Toldy, A., Szolnoki, B., & Marosi, G. (2011). Flame retardancy of fibre-reinforced epoxy resin composites for aerospace applications. *Polymer Degradation & Stability*, *96*(3), 371–376. doi:10.1016/j.polymdegradstab.2010.03.021

Tuncer, E., Sauers, I., James, D. R., Ellis, A. R., Paranthaman, M. P., Aytuğ, T., & Goyal, A. (2006). Electrical properties of epoxy resin based nano-composites. *Nanotechnology, 18*(2), 025703. doi:10.1088/0957-4484/18/2/025703

Utracki, L. A. (2004). *Clay-containing polymeric nanocomposites* (Vol. 1). iSmithers Rapra Publishing.

Wan, Y. J., Tang, L. C., Gong, L. X., Yan, D., Li, Y. B., Wu, L. B., ... Lai, G. Q. (2014). Grafting of epoxy chains onto graphene oxide for epoxy composites with improved mechanical and thermal properties. *Carbon, 69*, 467–480. doi:10.1016/j.carbon.2013.12.050

Wang, H., Hoa, S. V., & Wood-Adams, P. M. (2006). New method for the synthesis of clay/epoxy nanocomposites. *Journal of Applied Polymer Science, 100*(6), 4286–4296. doi:10.1002/app.23859

Wang, L., Wang, D., Dong, Z., Zhang, F., & Jin, J. (2013). Interface chemistry engineering for stable cycling of reduced GO/SnO2 nanocomposites for lithium ion battery. *Nano Letters, 13*(4), 1711–1716. doi:10.1021/nl400269d PMID:23477450

Wang, M., Fan, X., Thitsartarn, W., & He, C. (2015). Rheological and mechanical properties of epoxy/clay nanocomposites with enhanced tensile and fracture toughnesses. *Polymer, 58*, 43–52. doi:10.1016/j.polymer.2014.12.042

Wang, X. L., Yang, Y. Y., Chen, H. J., Wu, Y., & Ma, D. S. (2014). Synthesis of a vinylchlorine-containing 1, 3-diol from a marine cyanophyte. *Tetrahedron, 70*(30), 4571–4579. doi:10.1016/j.tet.2014.04.090

Wang, Z., Yang, X., Wang, Q., Hahn, H. T., Lee, S. G., Lee, K. H., & Guo, Z. (2011). Epoxy resin nanocomposites reinforced with ionized liquid stabilized carbon nanotubes. *International Journal of Smart and Nano Materials, 2*(3), 176–193. doi:10.1080/19475411.2011.594104

Wetzel, B., Rosso, P., Haupert, F., & Friedrich, K. (2006). Epoxy nanocomposites–fracture and toughening mechanisms. *Engineering Fracture Mechanics, 73*(16), 2375–2398. doi:10.1016/j.engfracmech.2006.05.018

Wu, F., Lu, Y., Shao, G., Zeng, F., & Wu, Q. (2012). Preparation of polyacrylonitrile/graphene oxide by in situ polymerization. *Polymer International, 61*(9), 1394–1399. doi:10.1002/pi.4221

Yeh, M. K., Hsieh, T. H., & Tai, N. H. (2008). Fabrication and mechanical properties of multi-walled carbon nanotubes/epoxy nanocomposites. *Materials Science and Engineering A, 483*, 289–292. doi:10.1016/j.msea.2006.09.138

Zhou, Y., White, E., Hosur, M., & Jeelani, S. (2010). Effect of particle size and weight fraction on the flexural strength and failure mode of TiO2 particles reinforced epoxy. *Materials Letters*, *64*(7), 806–809. doi:10.1016/j.matlet.2010.01.016

Zhu, J., Wei, S., Ryu, J., Budhathoki, M., Liang, G., & Guo, Z. (2010). In situ stabilized carbon nanofiber (CNF) reinforced epoxy nanocomposites. *Journal of Materials Chemistry*, *20*(23), 4937–4948. doi:10.1039/c0jm00063a

Chapter 8 Nanocomposites for Space Applications: A Review

Rafael Vargas-Bernal
https://orcid.org/0000-0003-4865-4575
Instituto Tecnológico Superior de Irapuato, Mexico

Margarita Tecpoyotl-Torres
https://orcid.org/0000-0002-4336-3771
Universidad Autónoma del Estado de Morelos, Mexico

ABSTRACT

A review on the advances achieved in the last 25 years in the development of hybrid nanocomposites based on polymer matrix for aerospace applications is presented here. The chapter analyzes the state-of-the-art strategies used in the design of materials that support the different conditions of the space environment. These materials are aimed primarily at structural applications, electromagnetic interference shielding, self-sensing, and self-healing, although they are not restricted to these applications. The introduction of metallic, ceramic, carbon-based nanomaterials such as carbon nanotubes and graphene, as well as two-dimensional materials have been used with a successful impact. Despite the significant advances that have been reached, much work must be done to achieve complete reliability for all properties required to protect the systems against the hazardous conditions found in space. Therefore, futuristic visions of the actions that must be carried out are raised in this chapter.

DOI: 10.4018/978-1-7998-1530-3.ch008

INTRODUCTION

Despite the technological advances achieved so far, scientists around the world are still looking for alternative materials that survive the conditions in space (Aïssa, 2019). Materials in space experience extreme conditions that are not found on Earth such as temperature, gravity, radiation, pressure, etc. The temperature in space ranges from -101 to 208 °C and during re-entry, the friction of the atmosphere will generate heat around the spacecraft up to 2760 °C. During the launch of spacecraft, they will experience up to three times the force of the Earth's gravity, and materials should not break, bend or weaken under such effects (Cander, 2019). Once the spacecraft reaches orbit, the force of gravity drops to almost zero, so this drastic change can affect the integrity of low-grade materials or those that do not support such efforts. In addition to this, materials in space are subject to radiation levels directly proportional to the height at which the space system orbits the planet. Therefore, these are regularly designed to operate in the low-earth orbit to experience less radiation than those operating in higher orbits or destined for even farther trips. Solar storms dramatically increase radiation levels without warning (Malandraki, 2018). Also, the materials in the spacecraft must withstand internal and external pressures, the first due to oxygen pressure of up to 15 pounds per square inch and the second due to gravitational changes. Due to the large amount of space debris produced by deceased satellites, spacecraft are exposed to impacts that can cause significant damage to their structure, so materials must be robust to withstand these events. The spontaneous presence of meteors with faster speeds than bullets (42 kilometers per second) can produce a hole in the weak parts of the structure. Finally, mechanical vibrations in a very wide range of frequencies must be avoided or minimized during and immediately after launch, so spacecraft materials should not present structural weaknesses. Therefore, each of the devices, systems or structures used in space must withstand similar conditions with varying levels of performance (Belous, 2017).

The materials that have been successfully used in the space environment are Kevlar (resistance to meteor and space debris impacts, durability, withstands extreme temperatures without damaging its structure or changing its shape), aluminum alloys (strength, light weight, resistance on impact), reinforced carbon-carbon composite coated with silicon carbide (withstands temperatures greater than 1260 °C), borosilicate glass (high and low temperature surface insulations), Nomex felt (surface insulation for temperatures below 371 °C in the middle and tail end), thermal glass (windows), silica (launch gears or loading site, low temperature zones), and sodium silicate (sealing of cracks). The use of polymer matrix nanocomposites allows the creation of optimal multifunctional materials for aerospace needs and applications (Rathod, 2017). Fillers used in this type of composite materials introduce new physical properties that polymers do not possess, resulting in high performance materials. The introduction of nanomaterials has led to improved capabilities by exploiting their mechanical, field emission, thermal, electrical and optical properties in aerospace design and applications. Future space missions require lightweight materials that maintain their physical properties for more than 30 years in hazardous environments such as exposure to atomic oxygen and solar radiation.

The space industry, like other technological areas of high degree of technological advancement such as the automotive industry, is looking for alternative materials to reduce fuel consumption and improving the safety and performance of the devices, systems or materials used in its practical applications (Rana, 2016). Composite materials offer a wide range of properties by combining fillers of different nature (polymers, ceramics, or metals), which are incorporated in different proportions, size (macro, micro or nano), shape (tubes, particles, short fibers, etc.), distribution and orientation (Balasubramanian, 2014; Rana, 2016). The introduction of nanocomposites has allowed many of the physical properties required of aerospace materials to be resolved favorably. Nanocomposites are the materials of twenty-first century having an annual growth rate of 25% due to their multifunctional capabilities (Rathod, 2017). In addition, these materials are expanding their research, design, and development for space applications and future space missions. For space applications, the main physical properties that must be considered are the mechanical, thermal, and electrical properties; while from the chemical point of view resistance or passivity to corrosion as well as biodegradability are of interest. The mechanical properties of space structures include strength, toughness, fatigue resistance, impact resistance and scratch resistance. At high altitudes, the space components also require properties such as low solar absorption, radiation resistance, high thermal emissivity and electrical conductivity. The proper choice of both the matrix and the filler is key to achieving the multifunctional properties required for many aerospace applications. Inorganic fillers for space applications such as fibers, whiskers, platelets and carbon nanotubes can contribute to improve mechanical strength, impact resistance, thermal stability and flame retardancy; while the polymer matrices have a low Young's modulus and shear strength, so they must be reinforced.

This chapter focuses mainly on advanced nanocomposites and their use in aerospace applications. The chapter has been divided as follows: The section called *Background* describes the main effects to which the components of space are subjected in the space environment. Next, in the section called, *Why use nanocomposites to protect space components?* the problems presented when coatings are not incorporated into the space component are analyzed. The role of nanocomposites in coatings is discussed in the section called *Solutions and recommendations*. A perspective on the progress that is planned to be achieved in the following decades is presented in the section called *Future research directions*. Finally, conclusions of the chapter are given.

BACKGROUND

Space materials for applications in low-Earth orbit (LEO) according to NASA are exposed to a unique environment: atomic oxygen (AO), ultraviolet and particulate radiation, meteorites and debris, vacuum, pollution, and thermal cycling (Young, 1993). Large fleets or constellations of communications satellites are scheduled for low-Earth orbits (LEO), which circulate the Earth at altitudes as low as 780 kilometers to as high as 1,300 kilometers, where multiple satellites follow each other in the same orbital path (Considine, 2008). Satellites with weights between 300 to 2,000 kg continue being launched to low-Earth orbits (LEO) to many scientific and Earth observation missions. Even the International Space Station (ISS) and MIR space station are found in low-Earth orbit (LEO). It is well knowing that a thin film of vitreous silica is used in space-based solar cells to protect to these against the damaging environment provided by atomic oxygen (AO), micrometeorites, and UV radiation effects found in space. Carbon nanotubes lead to be eroded by atomic oxygen (AO) in Earth's upper atmosphere (Considine, 2008). Through silent electric discharge by means of an electrical current to molecular oxygen (O_2), it is possible obtaining activated atomic oxygen (AO).

The relative abundance of atomic oxygen (AO) in the atmosphere begins to be appreciable at an altitude of 100 kilometers due to the photodissociation of oxygen by ultraviolet sunlight; above 200 kilometers, atomic oxygen (AO) is the main atmospheric constituent for several hundred kilometers (200 to 600 kilometers); and in addition, helium replaces atomic oxygen (AO) as the main constituent at altitudes between 600 and 1,500 kilometers according with sunspot cycle. Finally, at still higher altitudes, atomic hydrogen displaces helium as the main constituent. Hydrogen is present in a range of many radii of the Earth outside towards space, and constitutes the telluric hydrogen corona, or geocorona (Considine, 2008).

Space systems experience a wide range of conditions ranging from gaseous environments to vacuum, radiation from vacuum ultraviolet (VUV) to ionizing and passing through different debris environments (Chambers, 2010). The space environment is subjected to space weather caused by the changing activity of the sun. The activity of the space environment increases with the number of solar spots and flares which in turn increases the radiation emitted. Regularly, a cycle of 11 years elapses for a change of maximum activity to the minimum occurs. The atmospheric pressure in space operates in the range of 10^{-15} to 10^{-11} Pa, which can lead to sublimation or outgassing of some materials. For polymers used in thermal insulation and control, the rates of evaporation and/or loss can be significantly higher when the contained volatiles degas and modify the properties of the surface where they are condensed. Erosion by atomic oxygen (AO) in polymer could tear and fail very thin materials or make them rough because of the attack. Polymer erosion rates

Figure 1. Atomic-oxygen reactivity of different materials used in aerospace applications

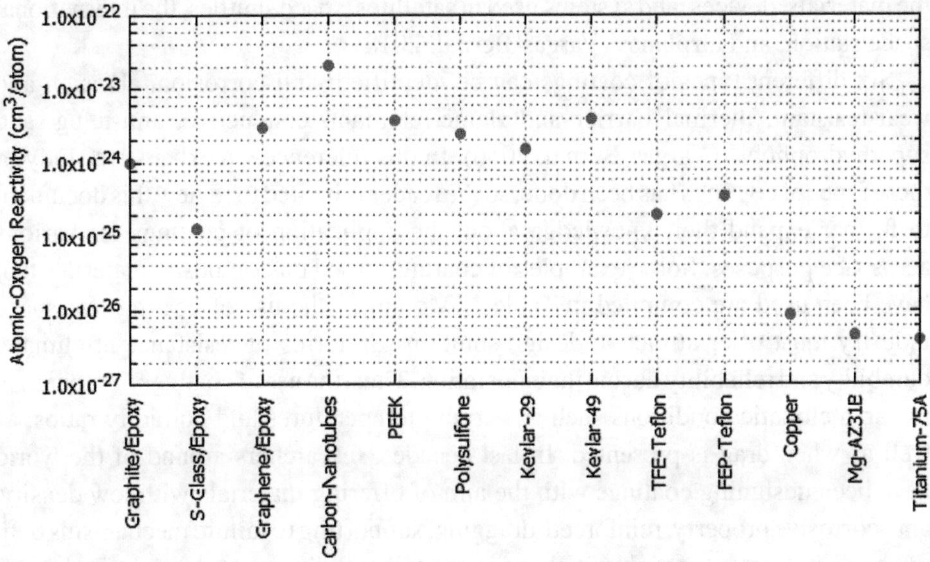

are variable and dependent on the composition and structure of the material. A graph of the atomic oxygen reactivity on different materials is shown in Figure 1. Metals have the lowest reactivity, composite materials have an intermediate reactivity, and polymers regularly have the highest reactivity. The data to obtain the graph were taken from the following references: (NASA, 1994), (Reddy, 1995), (Zhang, 2013), and (Liu, 2018).

The following section analyzes why researchers around the world propose nanocomposites as materials to solve the problems of the space environment.

WHY USE NANOCOMPOSITES TO PROTECT SPACE COMPONENTS?

Currently, space systems are suffering high maintenance cost and poor availability of assets. To reduce these problems, materials researchers have developed a set of coatings to overcome the different aspects where a reduction of physical and chemical properties has been presented. These materials present premature damages due to deterioration produced by hot erosion, distortion, clearance, and corrosion, as well as, by impact failure during operating cycles. Surfaces of aerospace components are subjected to heat, water, ice, electromagnetic waves, corrosion, abrasion, wear, thermal shock and/or fire during their lifetime (Vargas-Bernal, 2016). Aerospace applications experience electromagnetic interference produced by the space environment and by the materials, devices, and systems used in satellites, space shuttles, the international space station, and airplanes (Vargas-Bernal, 2018).

Six different types of coatings can be identified: anti-corrosion, abrasion and wear-resistant, thermal barrier and flame-retardant, conductive, anti-icing, and superhydrophobic (Vargas-Bernal, 2016). In this reference, an exhaustive study of these types of coatings has been done, so the reader is invited to review this document to further expand their knowledge about the application of coatings for various aerospace purposes. Some examples of coatings based on composite materials that have been used are compiled in Table 1. Moreover, the use of coatings for space industry must offer attractive design solutions, in terms of resistance at climate, durability and reliability during their operation. Thus, the use of coatings is imperative in harsh climatic conditions such as varying temperatures and humidity ratios, as well as when drag is presented. In last decades, researchers around of the world have been designing coatings with the aim of offering materials with low density, anti-corrosive property, reinforced-dragging, supporting to failure mechanisms both dynamic and static type, abrupt changes in harsh conditions, and good tribological properties. Materials science and engineering have introduced the use of coatings to protect the surface of the materials under these operating conditions. Materials such as ceramics, polymers, and metals and their combinations either as composite or hybrid materials, as well as functionalized versions of them, have been proposed as different design alternatives. A lot of studies presented here, that are related with aerospace applications, have been realized ex situ (in Earth) to be tested before of be launched to the space (in situ). Few experimental results have been revealed to the public from different aerospace agencies after that these materials have been tested in space. Results under simulation of space environments are published by different authors and here these are analyzed and discussed.

Among the polymeric materials with aerospace applications are the polyimides, since they present excellent flexibility, good mechanical and thermal stability, high glass transition temperatures and light weight (Wang, 2018b). Unfortunately, these materials when applied in the low Earth orbit (LEO) at altitudes ranging from 200 to 700 Km are subjected to the action of atomic oxygen (AO) that have an energy of 4.2 to 4.5 eV, which gradually deteriorates the covalent bonds and molecular chains of the polymer until their degradation. This event creates dangerous conditions for space systems which seriously threaten their security. Until now, atomic oxygen resistant coatings have been developed which may contain polysiloxanes, metal oxides, metal alloys, silica, diamond-like carbon, phenylphosphine oxides, polyhedral oligomeric silsesquioxanes (POSS), and other materials. The main problem of these

Table 1. Nanocomposite coatings with different aerospace applications

Type of Coating	Composite Materials Used	Reference(s)
Anti-wear	Tungsten carbide/Diamond-like carbon/Tungsten disulfide (WC/DLC/WS_2), Poly-ether-ketone/Graphene/Tungsten disulfide/Carbon nanotubes (PEEK/G/WS_2/CNTs), Chlorinated phenyl and methyl terminated silicone oil/polyimide (CPSO/PPI) or Perfluoropolyether/polyimide (PFPE/PPI)	(Voevodin, 1999) (Kalin, 2015) (Lv, 2015)
Electromagnetic interference (EMI) shielding	Graphene/Polystyrene (G/PS), Multi-wall carbon nanotubes/acrylonitrile butadiene styrene (MWCNT/ABS), Graphite/acrylonitrile butadiene styrene (C/ABS), Epoxy/carbonyl iron/cobalt-ferrite/calcium titanate (Epoxy/Carbonyl-Fe-$CoFe_2O_4$-$CaTiO_3$) Multi-wall carbon nanotubes/polystyrene (MWCNT/PS)	(Bera, 2015) (Afghahi, 2016) (Sachdev, 2016)
Anti-Icing	Epoxy/carbon nanotubes (Epoxy/CNTs), Poly(3,4-Ethylenedioxythiophene:Poly(styrenesulfonate)/Cyanoacrylate (PEDOT:PSS/Cyanoacrylate), Fluorinated silicone rubber/polyurethane (FSR/PU), Polytetrafluoroethylene/Graphene/Carbon nanotubes (PTFE/G/CNTs)	(Gohardani, 2013) (Matsubayashi, 2016) (Huang, 2019)
Anti-Corrosion	Polyaniline-Graphene oxide (PANI-GO), Epoxy/Polyaniline/Magnetite-Clay (Epoxy/PANI/Fe_3O_4/Clay), Nanoceria/Imidazole/Epoxy (CeO_2/Imidazole/Epoxy)	(Mooss, 2017) (Jlassi, 2018) (Motamedi, 2020)
Superhydrophobic	Chlorinated phenyl and methyl terminated silicone oil/polyimide (CPSO/PPI) or Perfluoropolyether/polyimide (PFPE/PPI), Polybutylene/Silica (PBI/SiO_2) or Polyethylene glycol/Silica (PEG/SiO_2), Fluorinated Polysiloxane-Methyltris(methylethylketoxime)silane-Titania (PMSF/MTO/TiO_2)	(Lv, 2015) (Wang, 2017) (Chen, 2018)
Thermal barrier	Bisphenol-A based Phtalonitrile Resin/Titania(TiO_2), Acrylic resin-Graphene oxide(GO), Polyimide/Al-YSZ or Cu-6Sn-YSZ or Cu-6Sn-NiCrAlY-YSZ	(Derradji, 2016) (Dong, 2016) (Abedi, 2018)
Flame retardant	Silica/Polypropylene (SiO_2/PP), Aluminium tri-hydroxide/High density polyethylene ($Al(OH)_3$-HDPE), Polyacrylic acid/Polyethylenimine/Carbon nanotubes (PAA/PEI/CNTs), Epoxy/Halloysite nanotubes	(Bar, 2015) (Qiu, 2018) (Vahadi, 2018)

coatings are the defects and the mechanical damage that are generated without the necessary care during their deposition.

The coatings applied to these spatial parts are also susceptible to cracks caused by events such as repeated cuts, punches, or bends. Openings and cracks can be produced on the coatings thanks to the inevitable collision of these materials with micrometeorites and space debris (Wang, 2018b). Space systems are also subjected to repeated thermal cycling that can cause cracking of the coatings because the surface temperature can range from 100 to -100 °C when the systems fly towards or opposite the sun. In addition to all these damages on the coatings, the mechanical stability is drastically reduced, and the service life is greatly reduced. Therefore, it is extremely important to design and manufacture coatings that are capable of autonomously healing the damage that they experience to significantly extend their service life under dangerous low Earth orbit (LEO) environment. In this way, the need for self-healing materials that may be capable of repairing mechanical damage and restoring their original functions is imperative to improve the reliability of space systems. Two major challenges must be overcome to manufacture reliable coatings: 1) the healing of the cracks requires an intimate contact of the polymers on the damaged surfaces, which implies that the polymer chains have a high mobility, and 2) the use of nanocomposites makes the mechanically robust coatings, which reduces the mobility of polymer chains.

Electromagnetic interference shielding is required in a wide range of applications such as communications and telecommunications systems, computers and data processing, aviation and aerospace systems, military and/or defense systems, automotive systems, office equipment, instrumentation and control systems, consumer electronics, audio and video systems, as well as in radio, television and recording systems (Sachdev, 2016). Civil or military metal structures require surface coatings that effectively absorb electromagnetic waves to avoid electromagnetic interference in a wide bandwidth, they must be have a small thickness and a light weight (Afghahi, 2016). Electromagnetic interference (EMI) is defined as an electrical perturbation that affects an electrical circuit, either due to electromagnetic induction or electromagnetic radiation emitted from an external source. The perturbation may interrupt, obstruct, or otherwise degrade or limit the effective performance of electrical circuits used in aerospace applications. These effects can range from a simple degradation of data to a total loss of data. This effect can be controlled by applying coatings with high electrical conductance or low electrical resistance (Vargas-Bernal, 2016). Nanocomposites represent a technological alternative to develop coatings that can offer a better shielding effectiveness against electromagnetic interference due to the possibility of controlling its electrical and magnetic properties as well as to that the size of the materials is very similar to the electromagnetic waves that it receives (Vargas-Bernal, 2018).

From a tribological point of view, the space systems are subjected to low lubrication and excessive wear. In addition, the materials remaining in space for a period of 10 to 30 years, are exposed to thermal cycling from cryogenic temperatures to values of 400 °C, energetic particles, solar radiation and atomic oxygen. Before being launched into space systems are tested on land and stored for many years under controlled environments (Voevodin, 1999). Coatings based on metals or ceramics reinforced with tungsten carbide (WC), tungsten disulphide (WS_2), diamond-like carbon (DLC), molybdenum disulfide (MOS_2) have been proposed to reduce tribological effects.

Novel nanocomposites based on the polyimide 2,4-bis(4-aminophenylamido-6-chloroquinazoline), pyromellitic dianhydride and functionalized-multiwalled carbon nanotubes (MWCNT) presented high glass transition temperature, high dielectric constant, and high flame retardancy (Govindaraj, 2018). The properties of this nanocomposite allow it to be used in microelectronics and space applications (T_g values from 338 °C to 375 °C, with dielectric constant value of 3.43, and limiting oxygen values in the range of 29.5 to 35.5). In addition, polyimides prepared using heterocyclic diamine monomers favor the tensile strength and Young's modulus.

Proper dispersion of nanoclay improves the properties of polymer matrix composites such as thermal stability, mechanical strength, gas barrier, and flame retardancy (Gul, 2017). In addition to improving these properties, its light weight and low cost have increased its demand in the space, automotive, barrier materials, construction and biomedical industries. Nanoclays are hydrophilic while polymers are hydrophobic. To increase compatibility between the two distinct phases, surface modification of nanoclays is achieved by treating them with quaternary alkyl ammonium salts. These salts can replace the interlayer small cations of nanoclays depending on the charge exchange capacity (CEC) of clay; as a result, inorganic nanoclay gets organically modified.

A space grade epoxy resin-clay nanocomposite requires that the clay be highly exfoliated with random and ordered orientations to produce an intercalated structure with considerable compression properties. In the case of nanocomposites based on epoxy resin and nanoclay, the presence of the latter improves the compressive strength and stiffness, as well as promotes the higher plastic hardening after manufacture and improves the fracture toughness of the polymer (Jumahat, 2012). The addition of mechanical fillers such as nanoscale clays to polymer resins produces several desirable effects such as: 1) increases Young's modulus without sacrificing strain to failure, 2) improving dimensional stability and thermal properties, 3) increasing the mechanical strength and yield stress, and 4) increases the resistance to the initiation and propagation of cracks.

Polymeric matrix nanocomposites for space applications such as polypropylene, polyaniline, polyurethane, polystyrene, and polyamide should be reinforced with carbon nanotubes, graphite, graphene oxide and clay to achieve attractive properties such as high glass transition temperature, thermal stability, high Young's modulus, chemical resistance and nonflammability (Kausar, 2017). Normally, composites based on organic matrix and carbon fibers are used in the aerospace and aircraft industry because they have high strength, Young's modulus, light weight, good fatigue life, high stiffness, and excellent corrosion resistance.

Organically modified superparamagnetic $MnFe_2O_4$/thermoplastic polyurethane elastomer (TPU) nanocomposites can be used for space deployable structures where a small compact, lightweight volume needs to undergo sudden large shape changes (Yoonessi, 2011). These materials can be used in the development of actuators activated by magnetostatics or electromagnetic fields.

Under conditions of icing in space systems, the deposition of supercooled liquid droplets on its surface is dependent on temperature, liquid water concentration and material properties (Gohardani, 2013). Epoxy resins reinforced with carbon nanotubes have low ice adhesion resistance, which allows them to be detached from the surfaces of the space components.

Nanocomposites based on polydimethylsiloxane (PDMS) with carbon nanotubes can be used as coatings for high energy proton shielding applications (53.6 MeV-105 MeV) in the aerospace industry (Li, 2013). High-energy radiation particles such as protons can induce damage to structural materials and electronic components of space systems. The proton radiation in the low Earth orbit has energies up to 250 MeV.

Radiation creates defects that impact the structure and electronic properties of materials (Walker II, 2016). Preliminary studies predict that two-dimensional materials can withstand high radiation environments found in space and produced by X-rays, gamma rays, electrons, protons, and heavy ions. Two-dimensional materials such as graphene, arsenene, aluminene, germanene, silicene, phosphorene, borophene, transition metal dichalcogenides (TMDs), Mxenes, etc. can provide these materials that withstand high energy radiation and thus reinforce the polymers in space.

The following subsection briefly describes the main types of synthesis techniques that can be used for the manufacture of polymeric nanocomposites.

Synthesis of Polymer Nanocomposites

Polymer nanocomposites can be defined as hybrid organic-inorganic materials where at least one dimension of the reinforcement phase is less than 100 nm (Fawaz, 2015). These are synthesized through different methods which can be classified as four main routes: melt intercalation, exfoliation adsorption, template synthesis, and in situ polymerization intercalation. In addition, according to the type of method and materials used in the synthesis, three different types of microstructure can be presented: unintercalated (or microcomposite), intercalated (and/or flocculated), or exfoliated (or also called delaminated). For example, recent advances in the synthesis of graphene-polymer nanocomposites were reported in (Zhang, 2015) and shown in Figure 2.

Melt intercalation, normally used for thermoplastic polymer matrices, involves annealing the matrix at high temperatures, adding the reinforcement, and finally mixing the composite material to achieve a uniform distribution. This method has

Figure 2. Examples of synthesis techniques of polymeric nanocomposites

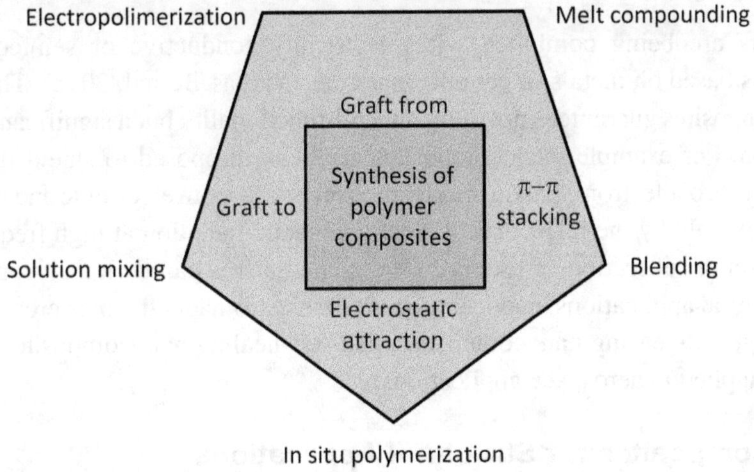

the advantages of being environmentally friendly thanks to the absence of solvents and compatible with industrial processes, while the disadvantage is the damage on the surface of the reinforcement due to high temperature processing.

Exfoliation adsorption involves working both constituents separately and then mixing them and polymerized: 1) the polymer must be dissolved through a solvent, and 2) the reinforcement must be swollen and dispersed in the same solvent. This technique is used when the polymer is soluble in water to produce intercalated nanocomposites. This is not environmentally friendly due to the use of solvents.

Template synthesis involves the formation of the composite material in an aqueous solution or gel containing the polymer and the reinforcement, where the polymer functions as a nucleating agent and promotes the growth of the reinforcement crystals and subsequent entrapment after polymerization.

The in-situ polymerization technique involves welling of the reinforcement in liquid monomer or monomer solution forming the interlayers and polymerization by heat, radiation, diffusion of the organic initiator or catalyst fixed through cation exchange. The advantage is that this approach produces better exfoliation than melt or exfoliation adsorption methods.

Some solutions and recommendations for the use of polymeric nanocomposites are analyzed below. These are aimed at presenting successful cases of their structural application, electromagnetic interference (EMI) shielding, as well as self-sensing and self-healing materials.

SOLUTIONS AND RECOMMENDATIONS

Polymers are being combined with electrically conductive or semiconductor materials based on metals or ceramic materials (Vargas-Bernal, 2018). The use of nanocomposites guarantees not only the multifunctionality but a significant weight reduction. For example, nanocomposites are being proposed to shield the space components of electromagnetic interference produced from electronic and radiation sources mainly by the reflection of electromagnetic radiation at high frequencies. Next, a set of subsections presents different alternatives based on nanocomposites for structural applications, nanocomposites for electromagnetic interference (EMI) shielding, self-sensing nanocomposites, and self-healing nanocomposites that are widely applied in aerospace applications.

Nanocomposites for Structural Applications

From the structural point-of-view, polymer matrix nanocomposites offer significant advantages since they can be used in the reinforcement and rehabilitation of damaged structures given the versatility of the type of nanoscale reinforcements that can be incorporated into polymers such as: carbon nanotubes, nano-clay, graphene, nanosilica, etc. Nanocomposites can be defined as those materials formed by a matrix and filler regularly in nanometric size that can be used to reinforce or improve the functionalities of the matrix material (Rana, 2016). Nanofillers present strategic advantages such as high surface-to-volume ratio, better interface, smaller defect size, and being used in small volumetric fractions to improve properties. However, three technical disadvantages have limited the application of polymer matrix nanocomposites in space applications: 1) their higher electrical resistance prevents electromagnetic shielding, 2) their lower thermal conductivity increases the load on deicing systems, and 3) they have low impact resistant, moisture absorption, and ageing with time (Rathod, 2017).

The need of manufacture light aerospace parts, with high mechanical properties, low cost and high operating temperatures has led to the use of aluminum, magnesium and titanium alloys. However, the demanding mechanical and thermal requirements have stimulated the research of the reinforcement of these alloys using other fillers. Graphene has been proposed as an alternative to improve the Young's Modulus ($\geq$ 165 GPa) and the thermal conductivity ($\geq$ 175 W/mK) through the formation of the magnesium-graphene nanocomposite as reported in (Das, 2018). This nanocomposite can be used in the manufacture of imperishable body parts and structures for aerospace applications as the authors point out.

Structural space applications can be developed using polyimide (PI) and clay, epoxy resin and carbon nanotubes (CNT), nylon and clay, epoxy resin and 10-dihydro-9-xa-phosphaphenanthrene-10-oxide (DOPO) based on phosphorus tetraglycidyl (PT) (Rathod, 2017). Deployable structures can be developed with composites based on epoxy resin and silicon carbide (SiC). Nanocomposites based on poly(phthalazionone ether sulfone ketone) and titanium dioxide (TiO_2) have tribological properties.

The tribological performance of nanocomposites based on poly-ether-ether-ketone (PEEK) and different nanoparticles: WS_2 fullerene-like (WS_2F), WS_2 needle-like (WS_2N), carbon nanotubes and graphene nanopowder, was studied in (Kalin, 2015). WS_2F improved the wear rate by 10% and WS_2N by 60%, while the carbon-based particles deteriorated the wear behavior by 20% and as much as three times in the case of the graphene. The addition of WS_2 fullerene-like nanoparticles and carbon nanotubes resulted in a lower friction coefficient compared to the effect of WS_2 needle-like nanoparticles and graphene. A graph containing the wear rate and friction coefficient values of the various polymeric nanocomposites described above is depicted in Figure 3. Wear rate values in the range of 10^{-7} to 10^{-4} mm^3/Nm are considered as mild wear or low wear rate caused by oxidative wear (Landolt, 2011).

Structural health monitoring can be performed through piezoelectric composites such as epoxy resin and lead zirconate titanate (PZT), poly(sodium 4-styrenesulfonate)-poly(vinyl alcohol) (PSS-PVA) and zinc oxide (ZnO); or through piezoresistive composites such as poly(methyl methacrylate) (PMMA) and carbon nanotubes (CNT) and carbon black (CB) (Rathod, 2017). Coatings based on nanocomposites can be manufactured using polyurethane (PU) and carbon nanotubes (CNT), polycarbonate (PC) and carbon nanotubes (CNT), epoxy resin and carbon nanotubes (CNT), epoxy resin and clay, epoxy resin and alumina (Al_2O_3), and poly(vinylidene fluoride) and carbon nanotubes (CNT) (Rathod, 2017). Space structural applications include coatings, aircraft interiors, housing, shrouds, equipment enclosures, cockpit, crew gear, durable space mirrors, nozzles, thermal shrinkage pipes, and solar array substrates.

Nanocomposites for Electromagnetic Interference (EMI) Shielding

With respect to electromagnetic shielding, polymer matrix nanocomposites offer significant advantages, since these can offer properties such as good electrical conductivity, corrosion resistance, low density and flexibility given the versatility of the type of nanoscale reinforcements that can be incorporated to polymers such as: carbon nanotubes, metal nanostructures, graphene, ceramic nanostructures, etc. Effective shielding to the electromagnetic interference of the coatings can be achieved when both electric fillers and magnetic filler are used (Afghahi, 2016). When only electrical fillers are used, a strong absorption with narrow bandwidth is

Figure 3. (a) Wear rate and (b) Coefficient of friction of different polymeric nanocomposites applied in aerospace applications

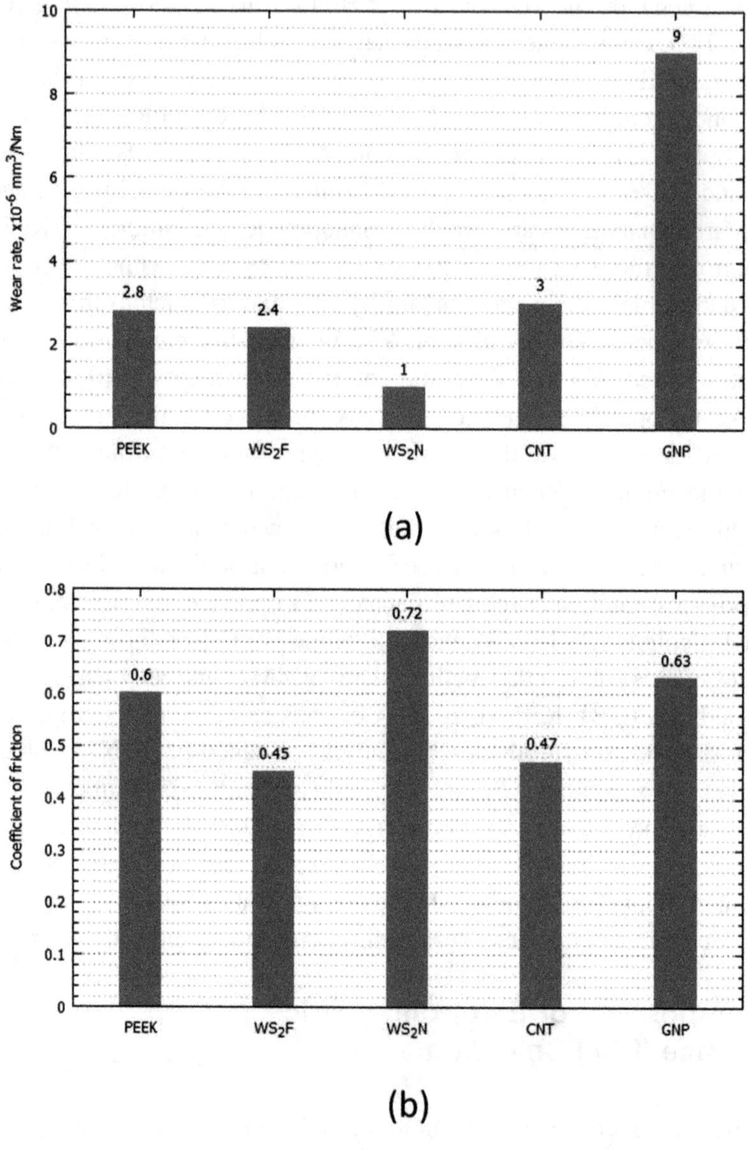

achieved, whereas when magnetic and electrical fillers are used together, slightly less absorption is obtained but in a wider bandwidth. The values of the electromagnetic absorption through the frequency range can be tuned through the thickness of the material used as coating. For example, epoxy resin nanocomposite with cobalt ferrite ($CoFeO_4$) (CF), calcium titanate ($CaTiO_3$) (CTO), and carbonyl iron (CI) powder produce an absorption of 33 dB in the X-band (8 to 12.4 GHz).

Nanocomposites based on acrylonitrile-butadiene-styrene (ABS) and multiwall carbon nanotubes (MWCNTs) were synthesized by (Sachdev, 2016), and these materials presented a better electromagnetic interference (EMI) shielding considering the reflection and absorption effects with the increment in volume fraction of carbon nanotubes. Polymeric matrix composites are considered versatile for this application because they are easy to synthesize, they are lightweight, they have an effective cost in large production, and mechanical properties such as strength and flexibility, as well as environmental resistance. Nanocomposites for electromagnetic interference shielding can be manufactured using epoxy resin and carbon nanotubes (CNT) or epoxy resin with fibers and carbon nanotubes (Rathod, 2017).

Carbon nanohorn (CNH)/graphene nanoplates (GNP)/poly(methyl methacrylate) (PMMA) nanocomposites present high electrical conductivity and electromagnetic interference shielding (Bera, 2015). These properties make carbon-based nanomaterials an interesting option for the development of defense and aerospace applications where shielding against electromagnetic interference is essential.

The three-dimensional addition of graphene to polyimide improves up to 1 order of magnitude the thermal conductivity and by 10 orders of magnitude the electrical conductivity of the nanocomposite formed (Loeblein, 2015). In addition, simulated space environment test results have shown that this material can support gamma rays and that its outgassing properties are within the allowed standards.

A graph containing the values of electromagnetic attenuation or shielding effectiveness of the various polymeric nanocomposites described above is illustrated in Figure 4. According to the Federal Communications Commission (FCC), the value of effective electromagnetic shielding effectiveness is above 30 dB (NASA, 1997).

Shape memory polymers and their composite materials present very interesting thermal and electrical responses for space applications, since they can be useful in deployable solar panels, reflecting antennas and morphing wings (Liu, 2014). These polymers are economic, have light weight, low density, good manufacturability, high deformability, electrical conductivity, transparency and permeability, good biodegradability and glass transition temperature easily tunable with respect to shape memory alloys and shape memory ceramics.

Figure 4. Shielding effectiveness of different polymeric nanocomposites applied in aerospace applications

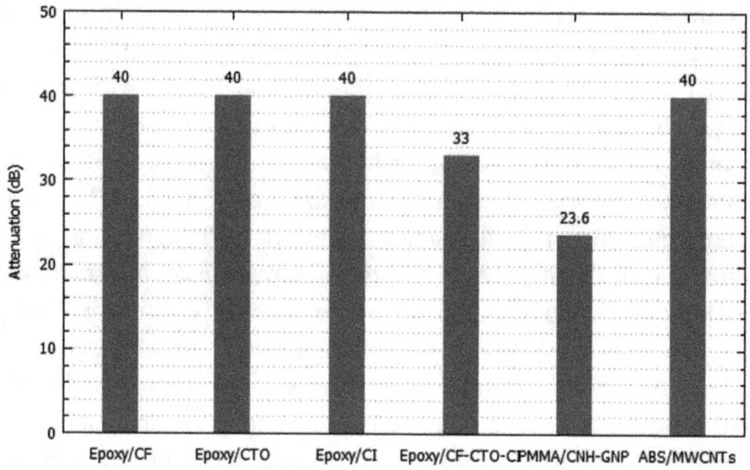

Self-Sensing Nanocomposites

Self-sensing nanocomposites can be defined as those materials that can sense their strain, damage, temperature, and other physical variables in order to increase their reliability from the point of view of safety (Rana, 2016). Before they were introduced, the monitoring was done through external sensors, however, they have a high cost, their installation requires highly trained personnel, they have a fragile nature, they are inaccessible in remote places, and present other disadvantages related to its instrumentation and communication with the control system. The principle of basic operation of these materials is piezoelectric, that is, they change their electrical resistance with strain or damage, which reduce the size of the electrical contact points between electrically conductive materials incorporated to perform the sensing, as shown in Figure 5. Among the materials that must be added to the nonconductive polymers to develop the self-sensing activity are carbon fibers, carbon particles or carbon nanotubes. Unfortunately, the filler materials must be highly sensitive to strain to produce considerable changes in electrical resistance before the complete breakdown of the composite material. Until now, carbon nanotubes have been able to detect microscale damage in structural composite materials with space applications (Avilés, 2018). The exhaustive use of nanomaterials in the various cutting-edge technological applications has led to the use of graphene to improve, in addition to the mechanical properties, to increase the electrical conductivity and the coupling of multifunctionality for the development of several electromechanical sensors such as those required in the self-sensing. To obtain an electrical behavior of the

Figure 5. A self-sensing nanocomposite based on piezoelectric materials

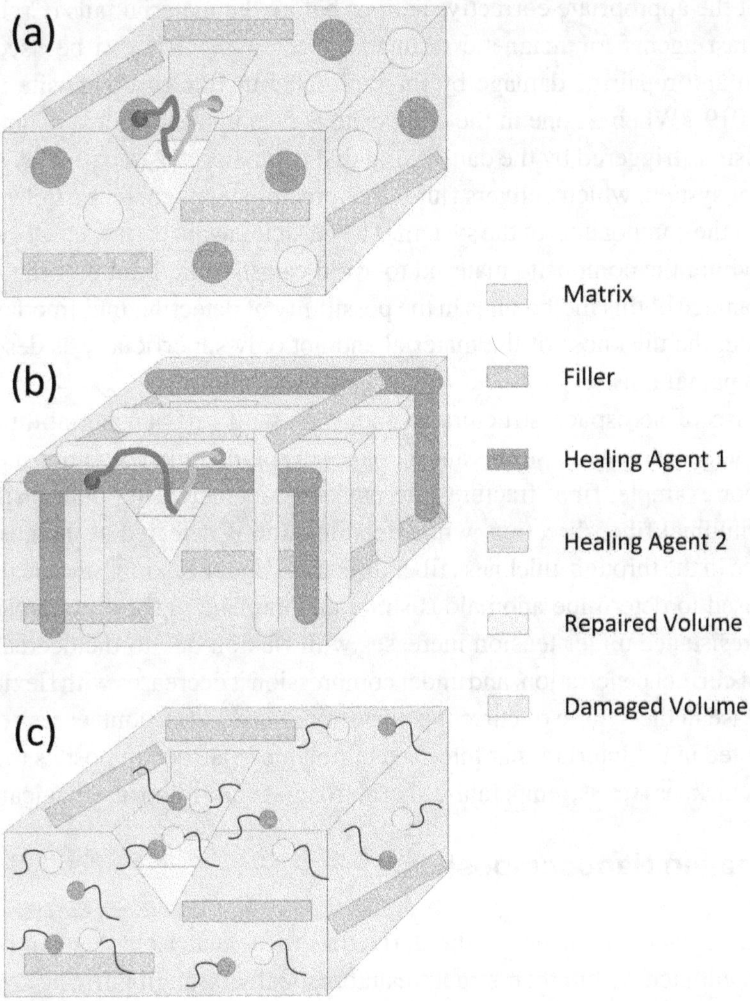

nanocomposite, the inclusion of fillers must be numerically enough to produce a percolation threshold, which can generate changes in electrical conductivity by means of strain, humidity, temperature and other external excitations. Thermosets polymers are used especially for the self-sensing of strain and damage in applications where load-bearing capacity is also required. The use of polymeric materials is focused on the development of sensors for human movement, soft skins, smart wearable sensors, and robotics, since in these applications large deformations are required.

Composite materials compared to metals have a reduced damage tolerance, and particularly fiber-reinforced materials are susceptible to impact damage that causes delamination which comprises the strength of the structure. Therefore, to guarantee the safety of the material, it is necessary to detect the presence of damage early to carry out the appropriate corrective actions before the material fails (Crall, 2019).

Contrast agents for magnetic resonance imaging (MRI) can be used for the detection and repair of damage by internal delamination of composite materials (Crall, 2019). When a zone in the composite is delaminated, then a magnetization mechanism is triggered by the damage, so that the material is repaired by means of a vascular system, which delivers fluids that precipitate magnetic particles after the mixing of the components of the system. The particles again connect the fibers to the matrix within the composite material to avoid catastrophic failure of the material. The advantage of this method lies in the possibility of detecting multimodal damage throughout the thickness of the material and not only superficially as described in previous paragraphs.

The use of aerospace structural composites with self-sensing abilities could allow to detect anomalous strains and damages in polymer matrix composites (Rana, 2016). For example, fiber fractures are presented as increasing bulk resistance in the longitudinal fiber direction, while delamination is detected as increasing bulk resistance in the through-thickness fiber direction. Under flexion, surface resistance can be used to determine anomalous strain and damage in these composites. The surface resistance under tension increases with flexion due to the decrease in the degree of current penetration, and under compression it decreases with flexion due to the increase in the degree of current penetration. In addition, contact resistivity can be exploited in the interlaminar interface of polymer matrix composites to measure through-thickness stress, temperature, thermal damage and minor mechanical damage.

Self-Healing Nanocomposites

Self-healing nanocomposite can be defined as those materials that can repair the damages initiated within their structure automatically through intrinsic or extrinsic mechanisms (Rana, 2016). Although, these mechanisms can be theoretically used in all types of materials, polymer matrices are mainly benefited with the scientific advances achieved until now. Therefore, in this section, the advantages offered by self-healing are directed to polymers used as matrices in composite materials and fillers present better properties that regularly are not healed. In self-healing polymer matrix nanocomposites, reinforcements provide convenient functionalities and various chemicals are used as the healing mechanism that will be triggered by mechanical damage and/or rupture (Thakur, 2015). Through the self-healing process, it is possible to repair from 75 to 100% the different mechanical properties of a nanocomposite material with respect to the same pristine material. The intrinsic mechanism implies that the polymeric material used in the matrix of the nanocomposite material can exhibit reversible interactions due to the action of external stimuli such as heat, light, radiation or the like. Once the damage is initiated, external stimuli can produce different reversible phase changes in thermoplastics, reversible chemical bonding as retro-Diels-Alder reactions (rDA), reversible physical interactions in ionomeric polymers, shape memory interactions, etc., (as illustrated in Figure 6) and these are responsible for repairing the cracks (Rana, 2016). In the case of a reversible phase change, the material is stimulated by energy to change its phase and it can return to its original form when this energy is released. Retro-Diels-Alder reaction is the dissociation wherein the Diels-Alder (DA) adduct reverts into its original components. Ionomeric polymers are capable of make (with stimulus) and breaking (without stimulus) the bonds in its structure. Shape memory polymers represents materials that can recover their permanent shapes thanks to the exposure to inverse stimuli that they undergone. In the case of the extrinsic mechanism, nanocomposites must contain an additional healing agent to fix the damage, which is located in microcapsules, hollow fibers or vascular networks, this is released when the cracks are produced, and the latter are repaired. The healing efficiency of the composite material can be determined through the equation

$$\eta = \frac{f_{healed} - f_{damaged}}{f_{virgin} - f_{damaged}} \tag{1}$$

where f is the property of interest in the virgin, damaged and healed states of the material (Blaiszik, 2010). The properties that can be repaired in self-healing composites are fracture, fatigue, impact resistance, corrosion, and barrier properties. Nanomaterials such as carbon nanotubes, graphene, nanoclay, nanocellulose, silica nanoparticles, halloysite nanotubes, silver nanoparticles, and gold nanoparticles have been used as reinforcements in self-healing polymer nanocomposites (Thakur, 2015). The encapsulation of liquid self-healing monomers is based on polyacrylonitrile (PAN) fiber or carbon nanotubes. The monomers are regularly based on dicyclopentadiene (DCPD) and isophorone diisocyanate (IPDI).

In self-healing polymers, the chemicals associated with repair are based on hydrogen bonding, π-π interactions, ionomers, and coordination bonds (Thakur, 2015). Self-healing mechanism can repair and recover functionality of the material using the resources inherently available in it, exploiting those components included from its manufacture. The presence of any damage in the material triggers the recovery process. All classes of polymers, that is, thermoplastics, thermosets and elastomers have potential for self-healing. Self-sealing systems are composed by three steps (Blaiszik, 2010). The first step consists in triggering or actuating the mechanism, which is closely coupled to the timescale of produced damage. In the second step, the transport of materials to the site of damage in realized at a relatively rapid rate.

Figure 6. Effects of the external stimuli in self-healing of nanocomposites: (a) reversible phase changes, (b) retro-Diels-Alder reactions (rDA), (c) reversible physical interactions and (d) shape memory interactions

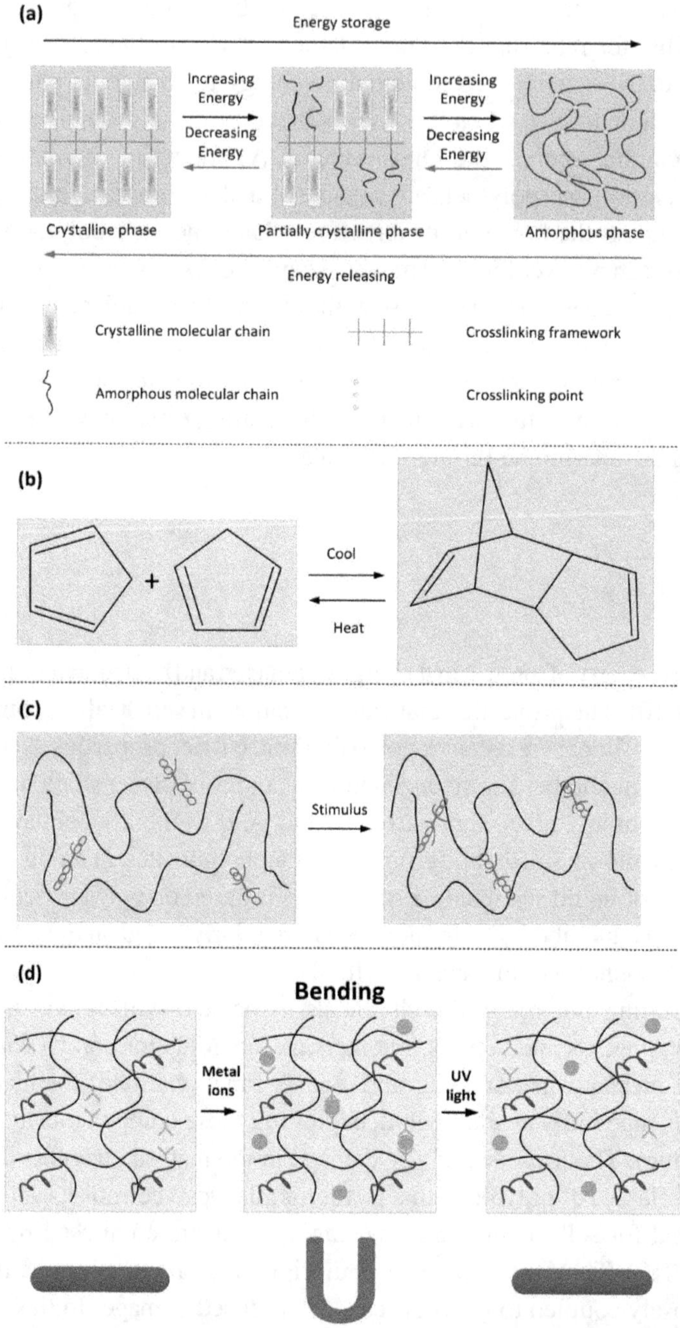

And finally, the chemical repair process developed at a timescale that is dependent of the type of healing required (for example, polymerization, entanglement, reversible cross-linking).

Traditionally, three different approaches can be implemented to make the healing namely, capsule-based healing, vascular healing and intrinsic healing, as shown in Figure 7. Capsule-based healing consists in the use of healing agents sequestered in discrete capsules. The capsule shell material is based on poly(urea-formaldehyde) (PUF), polyurethane (PU), poly(melamine-urea-formaldehyde) (PMUF), poly(melamine-formaldehyde) (PMF), etc. (Thakur, 2015). Vascular healing implies the use of healing agents delivered to the damage by an embedded vascular network having a 1D, 2D, or 3D connectivity and that can be replenished. Intrinsic healing uses a latent material ability triggered by damage through heat, light, or pressure. During damage an excess volume or damaged volume is generated by material separation which will be filled during the event of healing (Blaiszik, 2010). Capsule-based healing involves that discrete capsules are ruptured by damage, which release the healing agent and a chemical reaction is realized between composite and the agent until depleting the agent. Vascular healing uses a network in the form of capillaries connected either two or three dimensionally containing the healing agent which is activated when damage occurs. The great advantage of this approach consists in the possibility that these capillaries can be refilled by an external source or from an undamaged but connected region of the vascular system giving opportunity of developing multiple local healing events. Intrinsic healing implies reversible reactions in the polymer which rely on chain mobility and entanglement, reversible polymerizations, melting of thermoplastic phases, hydrogen bonding, or ionic interactions. Besides for this approach, there is possibility that multiple healing events can be realized. A more exhaustive study about self-healing is found in (Blaiszik, 2010).

Figure 7. Approaches to self-healing of nanocomposites: (a) capsule-based healing, (b) vascular healing, and (c) intrinsic healing

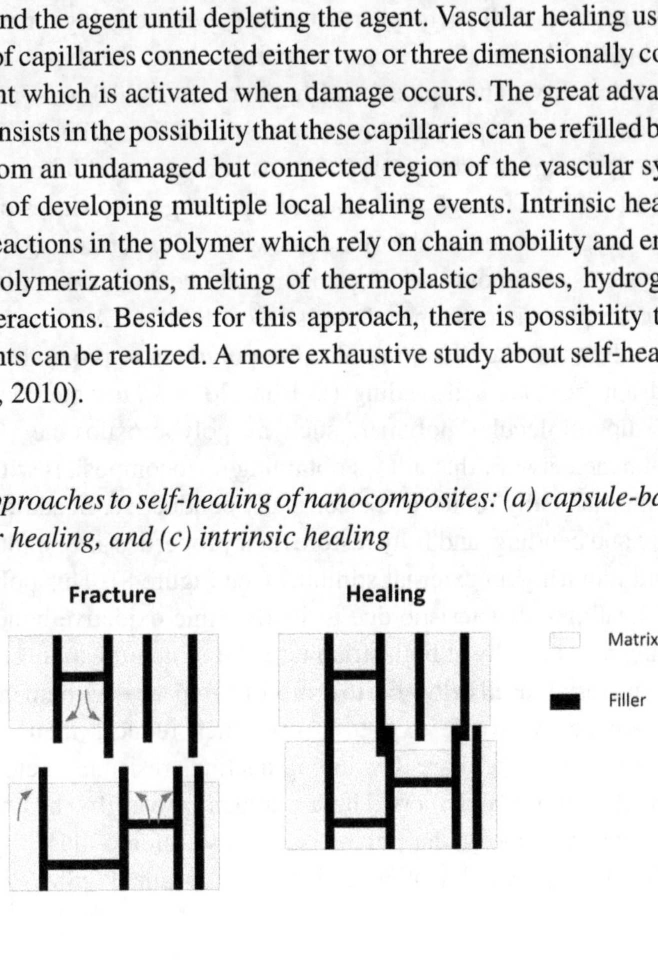

Recently, new polymeric materials that may indicate damage or exposure to excessive stress using a perceptible optical signal have been introduced (Calvino 2018). These materials, called mechanochromic or self-reporting polymers, modify their absorption and/or fluorescence properties with the application of a mechanical force. To perform this function, these materials incorporate microcapsules in the polymer used as matrix for nanocomposites that are filled with solutions of dyes that are released and chemically or physically activated, when the protective layer is damaged. In this way, it is possible to monitor structural health in materials with aeronautical, automotive and construction applications. In addition, motifs, called mechanophores, that contain mechanically reversible or irreversible labile bonds can be added to convert the mechanical forces into chemical reactions, which if they are mechanochromic will produce optical changes when these are activated by heat or light.

The possibility of developing synthetic materials that realize self-reporting of early damage to the material and self-healing simultaneously, based on biological principles, has been explored by (Wang, 2018a). This duality can be achieved by using pH sensitive core-shell microspheres used in polymer coatings on metals against corrosion, which will give them smart properties, long durability, recyclability, and multifunctionality. The spheres with nanometer thickness are based on poly(divinylbenzene)-graft-poly (divinylbenzene-co-methacrylic acid (PDVB-graft-P(DVB-co-AA)) contain benzotriazole (BTA) as a corrosion inhibitor and coumarin as a fluorescent probe. Through this polymeric coating the surfaces of metals can be exposed to oxygen, water, or chlorine ions, without experiencing damage.

Healing systems commonly develop incomplete healing, cannot heal multiple times, or require prior knowledge of the site of damage. A recently developed alternative involves the use of supramolecular polymers to promote a repeatable, efficient and autonomous self-healing (D'Elia, 2015). This method consists of confining a supramolecular polymer, such as polyborosiloxane (PBS), in an ultralight graphene network that allows obtaining nanocomposites with self-repair capabilities, mechanically robust and electrically conductive. In addition, they can sense pressure and bending, and fully restore their properties after multiple damage events without requiring an external stimulus (see Figure 8). This polymer has an intrinsic self-healing characteristic due to its dynamic oxidative bonds with both boron and silicon, especially at high strain rates for structural applications.

Polymeric materials used in low Earth orbit (LEO) space environment experience mechanical damage by atomic oxygen (AO) which reduces their service time. Researchers around the world are developing coatings resistant to atomic oxygen, which can heal the exposed polymers. These coatings are based on hydrogen-bonded three-dimensional supramolecular polymers such as 2-ureido-4[1H]-pyrimidinone (UPy)-functionalized polyhedral oligomeric silsesquioxane (POSS) (Wang, 2018b).

Figure 8. Healing process in nanocomposite based on supramolecular polymer and graphene

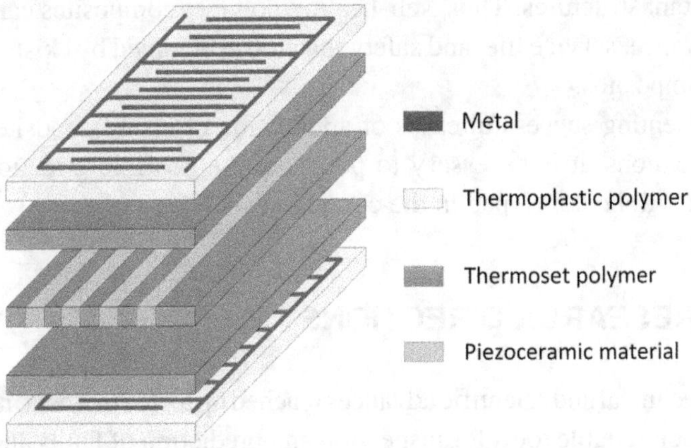

These polymers can be feasibly deposited on polyimides by a hot-pressing process thanks to the strong adhesion generated by the bending and/or unbending and thermal cycling treatments. The mechanism of resistance to attack by atomic oxygen is that in these coatings an epidermal SiO_2 layer is formed, which prevents the oxygen has contact with the coated part. In addition, these polymers are mechanically robust, thermally stable, and transparent.

A quite interesting and complete study of the healing techniques in found in (Cohades, 2018). The healing mechanism of composite materials is more complicated when a more refined synthesis method was used and when the number of components increases. In all cases, the control of the temperature during manufacturing and healing is important, since a balance must be maintained between the required mechanical performance and the desired embedded functionalities, because in some cases they can oppose themselves. Although in some cases, self-healing does not seem viable, this mechanism represents a desirable long-term objective to ensure the performance of materials for qualities as restrictive as space applications.

In aerospace applications, self-healing polymer composites are mainly applied in fuselages, aerostructures, and coatings (Rana, 2016). Fuselages based on fiber reinforced-polymers (FRPs) require self-healing to overcome impact load vulnerability, to protect structures from space debris, and to design low-maintenance, lightweight structures. Ionomeric polymers show excellent self-healing abilities under high-velocity impacts and good fragmentation abilities. Aircraft structural parts based on FRPs thanks to self-healing present a higher resistance to fatigue, due to that it heals microcracks before any crack growth or extension be presented. Additionally, self-healing coatings based on polymer composites could offer reduced recovery times after any damage such as corrosion and minor impacts. Future aerospace applications include anticorrosion coatings, smart paints, and impact-resistant structures. Thus, self-healing polymer composites can offer better damage tolerance, service life, and safety than those obtained by existing aerospace structural composites.

After presenting successful cases of application of nanocomposites in various spatial applications, it is necessary to point out the possible directions of future research that will be developed in the coming years.

FUTURE RESEARCH DIRECTIONS

Despite the technical and scientific advances reached up to this moment, it is necessary to develop soon reliable tools for inspection and prediction of faults and properties, as well as repairing techniques and standardization of composite materials. Within the mechanisms of inspection and prediction of faults, the technology of self-sensing composites represents one of the options for technological development and scientific research in the coming decades. Technologies such as self-curing must be studied and used more exhaustively in composite materials to reduce or eliminate the effects of external stimuli to which materials are subjected such as heat, light, radiation, etc., to increase the reliability of them. Understanding and applying the fundamental chemical mechanisms of space environmental effects on materials continues being the task of researchers around the world to develop more durable space materials.

Although the advances in the use of the self-sensing mechanism are significant, some important technical problems that have not been solved successfully must be resolved before the devices are taken to the commercialization stage, among them are the nonlinearity, cycling reproducibility, unwanted viscoelastic response, and hysteresis. There is the possibility of using hybrid materials based on carbon nanotubes and graphene or other two-dimensional materials to develop topological combinations to exploit both piezoimpedance and piezocapacitance in the sensing of strain and damage of polymer composites with self-sensing for space applications. In addition, numerous activities must be carried out to bring materials with self-sensing and self-healing to commercialization, among which are reliability studies and validation methods to obtain consumer confidence.

Finally, the conclusions reached after analyzing and describing the need to use polymeric nanocomposites in space applications are provided.

CONCLUSION

Nanocomposites due to their numerous attractive physical properties such as light weight, high strength, good corrosion resistance, higher fatigue resistance, high strain-to-failure, and multifunctionality, are being and will continue to be used in the space sector in the next decades. The combination of organic materials and inorganic materials produces hybrid materials such as nanocomposites that offer a wide variety of space applications which exploit the multifunctionality of nanomaterials used with reinforcements. In structural applications, composite materials offer a promising alternative to metallic materials because they provide excellent strength and stiffness with considerable weight reduction. The application of nanomaterials for the implementation of several types of coatings, structural materials, electromagnetic interference (EMI) shielding, self-sensing materials and self-healing materials has been successfully applied as described in the chapter. Despite the progress made so far, this area of research has much to offer in the coming decades of this century.

ACKNOWLEDGMENT

The first author acknowledges the support provided by CONRICYT of the National Council of Science and Technology (CONACYT) from Mexico for the access to scientific articles, as well as the support of the Thematic Network of Nanotechnology and Nanoscience and Thematic Network of Space Science and Technology of CONACYT, Mexico. Also, he wishes to thank to your family for their support in carrying out this research.

REFERENCES

Abedi, H. R., Salehi, M., & Shafyei, A. (2018). Multi-layered Thermal Barrier Coatings on BMI Polyimide Matrix Composite. *Surface and Coatings Technology, 337*, 104–116. doi:10.1016/j.surfcoat.2018.01.007

Afghahi, S. S. S., Jafarian, M., & Stergiou, C. A. (2016). Multicomponent Nanocomposites with Carbonyl Fe-$CoFe_2O_4$-$CaTiO_3$ Fillers for Microwave Absorption Applications. *Materials & Design, 112*, 462–468. doi:10.1016/j.matdes.2016.09.106

Aïssa, B., Haddad, E., & Jamroz, W. R. (Eds.). (2019). *Self-Healing Materials: From Fundamental Concepts to Advanced Space and Electronics Applications* (2nd ed.). London, UK: IET. doi:10.1049/PBCS070

Avilés, F., Oliva-Avilés, A. I., & Cen-Puc, M. (2018). Piezoresistivity, Strain, and Damage Self-Sensing of Polymer Composites Filled with Carbon Nanostructures. *Advanced Engineering Materials*, *20*(7), 1701159. doi:10.1002/adem.201701159

Balasubramanian, M. (2014). *Composite Materials and Processing*. Boca Raton, FL: CRC Press.

Bar, M., Alagirusamy, R., & Das, A. (2015). Flame Retardant Polymer Composites. *Fibers and Polymers*, *16*(4), 705–717. doi:10.100712221-015-0705-6

Belous, A., Saladukha, V., & Shvedau, S. (2017). Space Microelectronics: Vol. 1. *Modern Spacecraft Classification, Failure, and Electrical Component Requirements*. Norwood, MA: Artech House.

Bera, R., Suin, S., Maiti, S., Shrivastava, K., & Khatua, B. B. (2015). Carbon Nanohorn and Graphene Nanoplate based Polystyrene Nanocomposites for Superior Electromagnetic Interference Shielding Applications. *Journal of Applied Polymer Science*, *132*(46), 42803. doi:10.1002/app.42803

Blaiszik, B. J., Kramer, S. L. B., Olugebefola, S. C., Moore, J. S., Sottos, N. R., & White, S. R. (2010). Self-Healing Polymers and Composites. *Annual Review of Materials Research*, *40*(1), 179–211. doi:10.1146/annurev-matsci-070909-104532

Calvino, C., Weder, C. (2018). Microcapsule-Containing Self-Reporting Polymers. *Small*, *14*(46). DOI: doi:10.1002mll.201802489

Cander, L. R. (2019). *Ionospheric Space Weather*. Springer. doi:10.1007/978-3-319-99331-7

Chambers, A. R. (2010). Materials in Space: An Introduction and Overview. In R. Blockley & W. Shyy (Eds.), *Encyclopedia of Aerospace Engineering*. John Wiley & Sons. doi:10.1002/9780470686652.eae235

Chen, K., Gou, W., Xu, L., & Zhao, Y. (2018). Low Cost and Facile Preparation of Robust Multifunctional Coatings with Self-Healing Superhydrophobicity and High Conductivity. *Composites Science and Technology*, *156*, 177–185. doi:10.1016/j.compscitech.2017.12.036

Cohades, A., Branfoot, C., Rae, S., Bond, I., & Michaud, V. (2018). Progress in Self-Healing Fiber-Reinforced Polymer Composites. *Advanced Materials Interfaces*, *5*(17), 1800177. doi:10.1002/admi.201800177

Considine, G. D. (2008). *Van Nostrand's Scientific Encyclopedia* (10th ed.). John Wiley & Sons.

Crall, M. D., Laney, S. G., & Keller, M. W. (2019). Multimodal Damage Detection in Self-Sensing Fiber Reinforced Composites. *Advanced Functional Materials*, *29*(12), 1806634. doi:10.1002/adfm.201806634

D'Elia, E., Barg, S., Ni, N., Rocha, V. G., & Saíz, E. (2015). Self-Healing Graphene-based Composites with Sensing Capabilities. *Advanced Materials*, *27*(32), 4788–4794. doi:10.1002/adma.201501653 PMID:26178801

Das, D.K., & Sarkar, J. (2018). Graphene-Magnesium Nanocomposite: An Advanced Material for Aerospace Application. *Modern Physics Letters B*, *32*(6). DOI: doi:10.1142/S0217984918500756

Derradji, M., Ramdani, N., Zhang, T., Wang, J., Gong, L.-D., Xu, X.-D., ... Liu, W.-B. (2016). Effect of Silane Surface modified Titania Nanoparticles on the Thermal, Mechanical, and Corrosion Protective Properties of a Bisphenol-A based Phthalonitrile Resin. *Progress in Organic Coatings*, *90*, 34–43. doi:10.1016/j.porgcoat.2015.09.021

Dong, R., & Liu, L. (2016). Preparation and Properties of Acrylic Resin Coating modified by Functional Graphene Oxide. *Applied Surface Science*, *368*, 378–387. doi:10.1016/j.apsusc.2016.01.275

Fawaz, J., & Mittal, V. (2015). Synthesis of Polymer Nanocomposites: Review of Various Techniques. In V. Mittal (Ed.), *Synthesis Techniques for Polymer Nanocomposites* (pp. 1–30). Weinheim, Germany: Wiley-VCH. doi:10.1002/9783527670307.ch1

Gohardani, O., & Hammond, D. W. (2013). Ice Adhesion to Pristine and Eroded Polymer Matrix Composites Reinforced with Carbon Nanotubes for Potential Usage on Future Aircraft. *Cold Regions Science and Technology*, *96*, 8–16. doi:10.1016/j.coldregions.2013.09.005

Govindaraj, B., & Sarajadevi, M. (2018). Microwave-assisted Synthesis of Nanocomposites from Polyimides Chemically Cross-linked with Functionalized Carbon Nanotubes for Aerospace Applications. *Polymers for Advanced Technologies*, *29*(6), 1718–1726. doi:10.1002/pat.4275

Gul, S., Kausar, A., Muhammad, B., & Jabeen, S. (2017). Research Progress on Properties and Applications of Polymer/Clay Nanocomposite. *Polymer-Plastics Technology and Engineering*, *55*(7), 684–703. doi:10.1080/03602559.2015.1098699

Huang, X., Tepylo, N., Pommier-Budinger, V., Budinger, M., Bonaccurso, E., Villedieu, P., & Bennani, L. (2019). A Survey of Icephobic Coatings and their Potential Use in a Hybrid Coating/Active Ice Protection System for Aerospace Applications. *Progress in Aerospace Sciences*, *105*, 74–97. doi:10.1016/j.paerosci.2019.01.002

Jlassi, K., Radwan, A. B., Sadasivuni, K. K., Mrlik, M., Abdullah, A. M., Chehimi, M. M., & Krupa, I. (2018). Anti-Corrosive and Oil Sensitive Coatings based on Epoxy/Polyaniline/Magnetite-Clay Composites through Diazonium Interfacial Chemistry. *Scientific Reports*, *8*(1), 13369. doi:10.103841598-018-31508-0 PMID:30190528

Jumahat, A., Soutis, C., Jones, F. R., & Hodzic, A. (2012). Compressive Behaviour of Nanoclay modified Aerospace Grade Epoxy Polymer. *Plastics, Rubber and Composites*, *41*(6), 225–232. doi:10.1179/1743289811Y.0000000028

Kalin, M., Zalaznik, M., & Novak, S. (2015). Wear and Friction Behaviour of Poly-Ether-Ketone (PEEK) filled with Graphene, WS_2 and CNT Nanoparticles. *Wear*, *332-333*, 855–862. doi:10.1016/j.wear.2014.12.036

Kausar, A., Rafique, I., & Muhammad, B. (2017). Aerospace Application of Polymer Nanocomposite with Carbon Nanotube, Graphite, Graphene Oxide, and Nanoclay. *Polymer-Plastics Technology and Engineering*, *56*(13), 1438–1456. doi:10.1080/03602559.2016.1276594

Landolt, D., & Mischler, S. (Eds.). (2011). *Tribocorrosion of Passive Metals and Coatings*. Cambridge, UK: Woodhead Publishing. doi:10.1533/9780857093738

Li, Z., Nambiar, S., Zheng, W., & Yeow, J. T. W. (2013). PDMS/Single-Walled Carbon Nanotube Composite for Proton Radiation Shielding in Space Applications. *Materials Letters*, *108*, 79–83. doi:10.1016/j.matlet.2013.06.030

Liu, Y., Du, H., Liu, L., & Leng, J. (2014). Shape Memory Polymers and Their Composites in Aerospace Applications: A Review. *Smart Materials and Structures*, *23*(2), 023001. doi:10.1088/0964-1726/23/2/023001

Liu, Y., Yang, X., Li, M., Zhang, Y., Zhao, C., & Qin, W. (2018). Erosion Effects of Atomic Oxygen on Carbon Nanotube Arrays in Different Incidence Direction. *Surface and Interface Analysis*, *50*(5), 592–598. doi:10.1002ia.6416

Loeblein, M., Bolker, A., Tsang, S. H., Atar, N., Uzan-Saguy, C., Verker, R., ... Teo, E. H. T. (2015). 3D Graphene-Infused Polyimide with Enhanced Electrothermal Performance for Long-Term Flexible Space Applications. *Small*, *11*(48), 6425–6434. doi:10.1002mll.201502670 PMID:26479496

Lv, M., Wang, C., Wang, Q., Wang, T. & Liang, Y. (2015). Highly Stable Tribological Performance and Hydrophobicity of Porous Polyimide Material filled with Lubricants in a Simulated Space Environment. *RSC Advances*, *5*(66). DOI: doi:10.1039/c5ra23240f

Malandraki, O. E., & Crosby, N. B. (Eds.). (2018). *Solar Particle Radiation Storms Forecasting and Analysis: The HESPERIA HORIZON 2020 Project and Beyond.* Springer. doi:10.1007/978-3-319-60051-2

Matsubayashi, T., Tenjimbayashi, M., Manabe, K., Komine, M., Navarrini, W., & Shiratori, S. (2016). Integrated Anti-Icing Property of Super-Repellency and Electrothermogenesis Exhibited by PEDOT:PSS/Cyanocrylate Composite Nanoparticles. *ACS Applied Materials & Interfaces*, *8*(36), 24212–24220. doi:10.1021/acsami.6b07844 PMID:27540638

Mooss, V. A., Bhopale, A. A., Deshpande, P. P., & Athawale, A. A. (2017). Graphene oxide-modified Polyaniline Pigment for Epoxy based Anti-Corrosion Coatings. *Chemical Papers*, *71*(8), 1515–1528. doi:10.100711696-017-0146-3

Motamedi, M., Ramezanzadeh, M., Ramezanzadeh, B., & Mahdavian, M. (2020). One-Pot Synthesis and Construction of a High Performance Metal-Organic Structured Nano Pigment based on Nanoceria decorated Cerium(III)-Imidazole network (NC/CIN) for Effective Epoxy Composite Coating Anti-Corrosion and Thermo-Mechanical Properties improvement. *Chemical Engineering Journal*, *382*, 122820. doi:10.1016/j.cej.2019.122820

NASA. (1994). *LDEF Materials Results for Spacecraft Applications*. NASA Conference Publication 3257.

NASA. (1997). Design Guidelines for Shielding Effectiveness, Current Carrying Capability, and the Enhancement of Conductivity of Composite Materials. NASA Contractor Report 4784.

Qiu, X., Li, Z., Li, X., & Zhang, Z. (2018). Flame Retardant Coatings prepared using Layer by Layer Assembly: A Review. *Chemical Engineering Journal*, *334*, 108–122. doi:10.1016/j.cej.2017.09.194

Rana, S., & Fangueiro, R. (Eds.). (2016). *Advanced Composite Materials for Aerospace Engineering: Processing, Properties and Applications*. Cambridge, MA: Woodhead Publishing.

Rathod, V. T., Kumar, J. S., & Jain, A. (2017). Polymer and Ceramic Nanocomposites for Aerospace Applications. *Applied Nanoscience*, *7*(8), 519–548. doi:10.100713204-017-0592-9

Reddy, M. R. (1995). Effect of Low Earth Orbit Atomic Oxygen on Spacecraft Materials. *Journal of Materials Science*, *30*(2), 281–307. doi:10.1007/BF00354389

Sachdev, V. K., Sharma, S. K., Tomar, M., Gupta, V., & Tandon, R. P. (2016). EMI Shielding of MWCNT/ABS Nanocomposites in Contrast to Graphite/ABS Composites and MWCNT/PS Nanocomposites. *RSC Advances*, *6*(51), 45049–45058. doi:10.1039/C6RA04200G

Thakur, V. K., & Kessler, M. R. (2015). Self-Healing Polymer Nanocomposite Materials: A Review. *Polymer*, *69*, 369–383. doi:10.1016/j.polymer.2015.04.086

Vahadi, H., Saeb, M. R., Formela, K., & Cuesta, J. M. L. (2018). Flame Retardant Epoxy/Halloysite Nanotubes Nanocomposite Coatings: Exploring Low-Concentration Threshold for Flammability compared to Expandable Graphite as Superior Fire Retardant. *Progress in Organic Coatings*, *119*, 8–14. doi:10.1016/j.porgcoat.2018.02.005

Vargas-Bernal, R. (2016). Advances in Functional Nanocoatings Applied in the Aerospace Industry. In A. Zuzuarregui & M. C. Morant-Miñana (Eds.), *Research Perspectives on Functional Micro- and Nanoscale Coatings* (pp. 318–340). Hershey, PA: IGI Global. doi:10.4018/978-1-5225-0066-7.ch012

Vargas-Bernal, R., Bermúdez-Reyes, B., & Tecpoyotl-Torres, M. (2018). Progress in Advanced Materials used in Electromagnetic Interference Shielding for Space Applications. In C. D. Nikolopoulos (Ed.), *Electromagnetic Compatibility for Space Systems Design* (pp. 284–313). Hershey, PA: IGI Global. doi:10.4018/978-1-5225-5415-8.ch008

Voevodin, A. A., O'Neill, J. P., & Zabinski, J. S. (1999). Nanocomposite Tribological Coatings for Aerospace Applications. *Surface and Coatings Technology*, *116-119*, 36–45. doi:10.1016/S0257-8972(99)00228-5

Walker, R. C. II, Shi, T., Silva, E. C., Jovanovic, I., & Robinson, J. A. (2016). Radiation Effects on Two-Dimensional Materials. *Physica Status Solidi. A, Applications and Materials Science*, *213*(12), 3065–3077. doi:10.1002/pssa.201600395

Wang, J.-P., Wang, J.-K., Zhou, Q., Li, Z., Han, Y., Song, Y., … Li, G.L. (2018a). Adaptive Polymeric Coatings with Self-Reporting and Self-Healing Dual Functions from Porous Core-Shell Nanostructures. *Macromolecular Materials and Engineering*, *303*(4). DOI: doi:10.1002/mame.201700616

Wang, X., Li, Y., Qian, Y., Qi, H., Li, J., & Sun, J. (2018b). Mechanically Robust Atomic Oxygen-Resistant Coatings Capable of Autonomously Healing Damage in Low Earth Orbit Space Environment. *Advanced Materials*, *30*(36). DOI: doi:10.1002/adma.201803854

Wang, Y., & Gong, X. (2017). Superhydrophobic Coatings with Periodic Ring Structured Patterns for Self-Cleaning and Oil-Water Separation. *Advanced Materials Interfaces*, *4*(16), 1700190. doi:10.1002/admi.201700190

Yoonessi, M., Peck, J. A., Bail, J. L., Rogers, R. B., Lerch, B. A., & Meador, M. A. (2011). Transparent Large-Strain Thermoplastic Polyurethane Magnetoactive Nanocomposites. *ACS Applied Materials & Interfaces*, *3*(7), 2686–2693. doi:10.1021/am200468t PMID:21710967

Young, P. R., & Slemp, W. S. (1993). Space Environment Effects on Selected Long Duration Exposure Facility Polymeric Materials. In E. Reichmanis, C. W. Frank, & J. H. O'Donnell (Eds.), *Irradiation of Polymeric Materials: Processes, Mechanisms, and Applications* (pp. 278–304). Washington, DC: American Chemical Society. doi:10.1021/bk-1993-0527.ch018

Zhang, M., Li, Y., Su, Z., & Wei, G. (2015). Recent Advances in the Synthesis and Applications of Graphene-Polymer Nanocomposites. *Polymer Chemistry*, *6*(34), 6107–6124. doi:10.1039/C5PY00777A

Zhang, W., Yi, M., Shen, Z., Zhao, X., Zhang, X. & Ma, S. (2013). Graphene-Reinforced Epoxy Resin with Atomic Oxygen Erosion Resistance. *Journal of Materials Science*, *48*(6), 2416-2423. DOI: doi:10.100710853-012-7028-4

ADDITIONAL READING

Atar, N., Grossman, E., Gouzman, I., Bolker, A., & Hanein, Y. (2014). Reinforced Carbon Nanotube as Electrically Conducting and Flexible Films for Space Applications. *ACS Applied Materials & Interfaces*, *6*(22), 20400–20407. doi:10.1021/am505811g PMID:25366559

Gergely, R.C.R., Rossol, M.N., Tsubaki, S., Wang, J., Sottos, N.R., White, S.R. (2017). A Microvascular System for the Autonomous Regeneration of Large Scale Damage in Polymeric Coatings. *Advanced Engineering Materials*, 19(11): 1700319 (1-10). DOI: doi:10.1002/adem.201700319

Gergely, R.C.R., Sottos, N.R., White, S.R. (2017). Regenerative Polymeric Coatings enabled by Pressure Responsive Surface Valves. *Advanced Engineering Materials*, 19(11): 1700308(1-5). DOI: doi:10.1002/adem.201700308

Li, W., Matthews, C. C., Yang, K., Odarczenko, M. T., White, S. R., Sottos, N. R., & White, S. R. (2016). Autonomous Indication of Mechanical Damage in Polymeric Coatings. *Advanced Materials*, *28*(11), 2189–2194. doi:10.1002/adma.201505214 PMID:26754020

Xia, X., Mazzeo, A.D., Zhong, Z., Weng, G.J. (2017). An X-Band Theory of Electromagnetic Interference Shielding for Graphene-Polymer Nanocomposites. *Journal of Applied Physics*, 122(2): 025104(1-12). DOI: doi:10.1063/1.4992074

KEY TERMS AND DEFINITIONS

Carbon Nanotube (CNT): Allotrope of carbon with a cylindrical hollow nanostructure with nanowalls formed by one-atom-thick sheets of carbon rolled at specific and discrete angles.

Graphene: Allotrope of carbon formed by a single layer of carbon atoms located in a hexagonal lattice.

Hybrid Material: Material composed of a mixture of inorganic and organic components.

Mxenes: Two-dimensional inorganic compound with layers of thickness of few atoms containing transition metal carbides, nitrides, or carbonitrides.

Nanocomposite: Multiphase solid material where at least one of its phases has one, two, or three dimensions in the range of nanometers.

Nanoparticle: Organic or inorganic particle between 1 and 100 nanometers in size with a surrounding interfacial layer.

Self-Healing: Mechanism found in those materials that can repair to themselves the damages initiated within their structure automatically through intrinsic or extrinsic means.

Self-Sensing: Mechanism found in those materials that can sense to themselves: strain, damage, temperature and other physical variables, in order to increase their reliability from the point of view of safety.

Two-Dimensional Material: Crystalline material consisting of a single layer of atoms or a single layer of two or more covalently bonding elements.

Related Readings

To continue IGI Global's long-standing tradition of advancing innovation through emerging research, please find below a compiled list of recommended IGI Global book chapters and journal articles in the areas of nanotechnology, coatings, and materials science. These related readings will provide additional information and guidance to further enrich your knowledge and assist you with your own research.

Agboola, O., Popoola, P. A., Sadiku, R., Sanni, S. E., Babatunde, D. E., Abatan, O. G., ... Fasiku, O. V. (2019). Polymer Nanocomposites for Advanced Automobile Applications. In N. Ramdani (Ed.), *Polymer Nanocomposites for Advanced Engineering and Military Applications* (pp. 96–130). Hershey, PA: IGI Global. doi:10.4018/978-1-5225-7838-3.ch004

Ashrafiyan, O., Saremi, M., Pakseresht, A., & Ghasali, E. (2018). Oxidation-Protective Coatings for Carbon-Carbon Composites. In A. Pakseresht (Ed.), *Production, Properties, and Applications of High Temperature Coatings* (pp. 429–446). Hershey, PA: IGI Global. doi:10.4018/978-1-5225-4194-3.ch016

Aslam, S., & Ahmad, M. (2019). Nanoparticles for Degradation of Organic Pollutants. In R. Nazir (Ed.), *Nanotechnology Applications in Environmental Engineering* (pp. 241–267). Hershey, PA: IGI Global. doi:10.4018/978-1-5225-5745-6.ch011

Azadi, M. (2018). Deposition of TiN, TiC, and DLC Coatings by PACVD. In A. Pakseresht (Ed.), *Production, Properties, and Applications of High Temperature Coatings* (pp. 381–402). Hershey, PA: IGI Global. doi:10.4018/978-1-5225-4194-3.ch014

Babatunde, D. E., Denwigwe, I. H., Babatunde, O. M., Agboola, O., & Akinsipe, G. D. (2019). Relevance of Chemically Functionalized Nano-Fillers and Modified Nanocomposite in Energy Systems. In N. Ramdani (Ed.), *Polymer Nanocomposites for Advanced Engineering and Military Applications* (pp. 10–65). Hershey, PA: IGI Global. doi:10.4018/978-1-5225-7838-3.ch002

Bhattacharyya, S., & Majumdar, P. S. (2018). An Overview on Peak Shape Method for Thermoluminescence Glow Curve Analysis: Application on Tremolite and Actinolite Glow Peaks. In R. Tiwari, V. Dubey, & S. Dhoble (Eds.), *Emerging Synthesis Techniques for Luminescent Materials* (pp. 26–52). Hershey, PA: IGI Global. doi:10.4018/978-1-5225-5170-6.ch002

Bolleddu, V., Racherla, V., & Bandyopadhyay, P. P. (2018). Microstructural and Tribological Characteristics of Air Plasma Sprayed Alumina-Titania Coatings. In A. Pakseresht (Ed.), *Production, Properties, and Applications of High Temperature Coatings* (pp. 268–298). Hershey, PA: IGI Global. doi:10.4018/978-1-5225-4194-3.ch011

Borah, R., & Kumar, A. (2019). Enhanced Cellular Activity on Conducting Polymer: Chitosan Nanocomposite Scaffolds for Tissue Engineering Application. In N. Ramdani (Ed.), *Polymer Nanocomposites for Advanced Engineering and Military Applications* (pp. 150–189). Hershey, PA: IGI Global. doi:10.4018/978-1-5225-7838-3.ch006

Chawla, R., Singhal, P., & Garg, A. K. (2018). Investigation of Structural, Morphological and Electro-Optical Behavior of GO/rGO. *International Journal of Surface Engineering and Interdisciplinary Materials Science*, *6*(1), 32–43. doi:10.4018/IJSEIMS.2018010103

Chopra, S., Sreya, S., Babhulkar, R. V., Halde, S. P., Deshmukh, K. A., & Peshwe, D. R. (2019). Cryogenic Treatment of Polymer/MWCNT Nano-Composites for Mechanical and Tribological Applications. In N. Ramdani (Ed.), *Nanotechnology in Aerospace and Structural Mechanics* (pp. 103–161). Hershey, PA: IGI Global. doi:10.4018/978-1-5225-7921-2.ch004

Darwish, M. (2019). Lightweight Nanocomposites Polymers for Shielding Application. In N. Ramdani (Ed.), *Nanotechnology in Aerospace and Structural Mechanics* (pp. 206–233). Hershey, PA: IGI Global. doi:10.4018/978-1-5225-7921-2.ch006

Di Girolamo, G. (2018). Current Challenges and Future Perspectives in the Field of Thermal Barrier Coatings. In A. Pakseresht (Ed.), *Production, Properties, and Applications of High Temperature Coatings* (pp. 25–59). Hershey, PA: IGI Global. doi:10.4018/978-1-5225-4194-3.ch002

Dutta, M. M. (2019). Characterization of Nanomaterials. In R. Nazir (Ed.), *Nanotechnology Applications in Environmental Engineering* (pp. 44–80). Hershey, PA: IGI Global. doi:10.4018/978-1-5225-5745-6.ch003

Elehinafe, F. B., & Ayeni, A. O. (2019). Processing of Polymer-Based Nanocomposites in Advanced Engineering and Military Application. In N. Ramdani (Ed.), *Polymer Nanocomposites for Advanced Engineering and Military Applications* (pp. 1–9). Hershey, PA: IGI Global. doi:10.4018/978-1-5225-7838-3.ch001

Figueira, R. B. (2018). Hybrid Sol-Gel Coatings: Erosion-Corrosion Protection. In A. Pakseresht (Ed.), *Production, Properties, and Applications of High Temperature Coatings* (pp. 334–380). Hershey, PA: IGI Global. doi:10.4018/978-1-5225-4194-3.ch013

Gebai, S. S., Hallal, A. M., & Hammoud, M. S. (2018). *Mechanical Properties of Natural Fiber Reinforced Polymers: Emerging Research and Opportunities* (pp. 1–228). Hershey, PA: IGI Global. doi:10.4018/978-1-5225-4837-9

Hassanzadeh, M., Saremi, M., Valefi, Z., & Pakseresht, A. H. (2018). Investigation on Improved-Durability Thermal Barrier Coatings: An Overview of Nanostructured, Multilayered, and Self-Healing TBCs. In A. Pakseresht (Ed.), *Production, Properties, and Applications of High Temperature Coatings* (pp. 60–78). Hershey, PA: IGI Global. doi:10.4018/978-1-5225-4194-3.ch003

Henao, J., & Sotelo, O. (2018). Surface Engineering at High Temperature: Thermal Cycling and Corrosion Resistance. In A. Pakseresht (Ed.), *Production, Properties, and Applications of High Temperature Coatings* (pp. 131–159). Hershey, PA: IGI Global. doi:10.4018/978-1-5225-4194-3.ch006

Horst, D. J., & Junior, P. P. (2019). 3D-Printed Conductive Filaments Based on Carbon Nanostructures Embedded in a Polymer Matrix: A Review. *International Journal of Applied Nanotechnology Research, 4*(1), 26–40. doi:10.4018/IJANR.2019010103

Irzhak, T. F., & Irzhak, V. I. (2019). Synthesis of Epoxy Nanocomposites. In N. Ramdani (Ed.), *Nanotechnology in Aerospace and Structural Mechanics* (pp. 34–79). Hershey, PA: IGI Global. doi:10.4018/978-1-5225-7921-2.ch002

Ismar, E., & Sarac, A. (2019). Carbon Nanomaterials: Carbon Nanotubes, Graphene, and Carbon Nanofibers. In N. Ramdani (Ed.), *Nanotechnology in Aerospace and Structural Mechanics* (pp. 1–33). Hershey, PA: IGI Global. doi:10.4018/978-1-5225-7921-2.ch001

Jucha, S. O., & Moskal, G. J. (2018). Thermal Barrier Coatings on the Base of Samarium Zirconates. In A. Pakseresht (Ed.), *Production, Properties, and Applications of High Temperature Coatings* (pp. 79–106). Hershey, PA: IGI Global. doi:10.4018/978-1-5225-4194-3.ch004

Katrenipadu, S. P., & Gurugubelli, S. N. (2018). Regression Modeling and Experimental Investigations on Ageing Behavior of Nano-Fly Ash Reinforced Al-10wt%Mg Alloy Matrix Composites. *International Journal of Surface Engineering and Interdisciplinary Materials Science, 6*(2), 36–49. doi:10.4018/IJSEIMS.2018070103

Kausar, A. (2019). Aeronautical Impact of Epoxy/Carbon Nanotube Nanocomposite. In N. Ramdani (Ed.), *Nanotechnology in Aerospace and Structural Mechanics* (pp. 80–102). Hershey, PA: IGI Global. doi:10.4018/978-1-5225-7921-2.ch003

Kausar, A. (2019). Scope of Polymer/Graphene Nanocomposite in Defense Relevance: Defense Application of Polymer/Graphene. In N. Ramdani (Ed.), *Polymer Nanocomposites for Advanced Engineering and Military Applications* (pp. 296–315). Hershey, PA: IGI Global. doi:10.4018/978-1-5225-7838-3.ch010

Keshavamurthy, R., Naveena, B. E., & Sekhar, N. (2018). Thermal Spray Coatings for Erosion-Corrosion Protection. In A. Pakseresht (Ed.), *Production, Properties, and Applications of High Temperature Coatings* (pp. 246–267). Hershey, PA: IGI Global. doi:10.4018/978-1-5225-4194-3.ch010

Khelfi, A. (2019). Toxicocinetic and Mechanisms of Action of Nanoparticles. In R. Nazir (Ed.), *Nanotechnology Applications in Environmental Engineering* (pp. 344–368). Hershey, PA: IGI Global. doi:10.4018/978-1-5225-5745-6.ch014

Kosti, S. (2019). Nanomaterials and Nanocomposites Thermal and Mechanical Properties Modelling. In N. Ramdani (Ed.), *Nanotechnology in Aerospace and Structural Mechanics* (pp. 234–256). Hershey, PA: IGI Global. doi:10.4018/978-1-5225-7921-2.ch007

Kumar, I. (2019). Simulation and Modeling of Nanotechnology Aircraft Using MATLAB. In N. Ramdani (Ed.), *Nanotechnology in Aerospace and Structural Mechanics* (pp. 257–290). Hershey, PA: IGI Global. doi:10.4018/978-1-5225-7921-2.ch008

Kumar, S., Krishnan, S., Samal, S. K., Mohanty, S., & Nayak, S. K. (2019). Polymer Nanocomposites Coating for Anticorrosion Application. In N. Ramdani (Ed.), *Polymer Nanocomposites for Advanced Engineering and Military Applications* (pp. 254–294). Hershey, PA: IGI Global. doi:10.4018/978-1-5225-7838-3.ch009

Kuz'min, V. E., Ognichenko, L. N., Sizochenko, N., Chapkin, V. A., Stelmakh, S. I., Shyrykalova, A. O., & Leszczynski, J. (2019). Combining Features of Metal Oxide Nanoparticles: Nano-QSAR for Cytotoxicity. *International Journal of Quantitative Structure-Property Relationships*, *4*(1), 28–40. doi:10.4018/IJQSPR.2019010103

Lahariya, V., & Ramrakhiani, M. (2018). Electroluminescence Principle and Novel Electroluminescent Materials. In R. Tiwari, V. Dubey, & S. Dhoble (Eds.), *Emerging Synthesis Techniques for Luminescent Materials* (pp. 415–438). Hershey, PA: IGI Global. doi:10.4018/978-1-5225-5170-6.ch011

Lateef, A., & Nazir, R. (2019). Application of Magnetic Nanomaterials for Water Treatment. In R. Nazir (Ed.), *Nanotechnology Applications in Environmental Engineering* (pp. 201–219). Hershey, PA: IGI Global. doi:10.4018/978-1-5225-5745-6.ch009

Majumdar, D. (2019). Polyaniline as Proficient Electrode Material for Supercapacitor Applications: PANI Nanocomposites for Supercapacitor Applications. In N. Ramdani (Ed.), *Polymer Nanocomposites for Advanced Engineering and Military Applications* (pp. 190–219). Hershey, PA: IGI Global. doi:10.4018/978-1-5225-7838-3.ch007

Majumdar, D. (2019). Polyaniline Nanocomposites: Innovative Materials for Supercapacitor Applications – PANI Nanocomposites for Supercapacitor Applications. In N. Ramdani (Ed.), *Polymer Nanocomposites for Advanced Engineering and Military Applications* (pp. 220–253). Hershey, PA: IGI Global. doi:10.4018/978-1-5225-7838-3.ch008

Mubarak, S. M. (2019). Nanotechnology for Air Remediation. In R. Nazir (Ed.), *Nanotechnology Applications in Environmental Engineering* (pp. 121–142). Hershey, PA: IGI Global. doi:10.4018/978-1-5225-5745-6.ch006

Naik, R., Revathi, V., Prashantha, S. C., & Nagabhushana, H. (2018). Synthesis of Magnesium Based Nanophosphors and Nanocomposites by Different Techniques: Silicates and Ferrites. In R. Tiwari, V. Dubey, & S. Dhoble (Eds.), *Emerging Synthesis Techniques for Luminescent Materials* (pp. 251–276). Hershey, PA: IGI Global. doi:10.4018/978-1-5225-5170-6.ch007

Nasreen, S., & Rafique, U. (2019). Nanotechnology for Water Environmental Application: Functionalized Silica Hybrids as Nano-Sorbents. In R. Nazir (Ed.), *Nanotechnology Applications in Environmental Engineering* (pp. 171–200). Hershey, PA: IGI Global. doi:10.4018/978-1-5225-5745-6.ch008

Nazir, R. (2019). Introduction to Environmental Nanotechnology: E-Nano. In R. Nazir (Ed.), *Nanotechnology Applications in Environmental Engineering* (pp. 1–27). Hershey, PA: IGI Global. doi:10.4018/978-1-5225-5745-6.ch001

Noman, D., & Owens, G. (2019). Wastewater Treatment and Role of Green Synthesized Metal Oxide Nanocomposites. In R. Nazir (Ed.), *Nanotechnology Applications in Environmental Engineering* (pp. 268–307). Hershey, PA: IGI Global. doi:10.4018/978-1-5225-5745-6.ch012

Nosheen, S. (2019). Nanomembrane Applications in Environmental Engineering. In R. Nazir (Ed.), *Nanotechnology Applications in Environmental Engineering* (pp. 103–120). Hershey, PA: IGI Global. doi:10.4018/978-1-5225-5745-6.ch005

Odetola, P. I., Popoola, P. A., & Oladijo, P. (2018). Thin Coating Deposition by Magnetron Sputtering. In A. Pakseresht (Ed.), *Production, Properties, and Applications of High Temperature Coatings* (pp. 403–428). Hershey, PA: IGI Global. doi:10.4018/978-1-5225-4194-3.ch015

Pandey, A., Tiwari, S. P., Dutta, V., & Kumar, V. (2018). Photon Upconversion in Lanthanide-Activated Inorganic Luminescent Materials. In R. Tiwari, V. Dubey, & S. Dhoble (Eds.), *Emerging Synthesis Techniques for Luminescent Materials* (pp. 86–116). Hershey, PA: IGI Global. doi:10.4018/978-1-5225-5170-6.ch004

Pandey, J. (2019). Recent Progresses in Membranes for Proton Exchange Membrane Fuel Cell (PEMFC) for Clean and Environmentally Friendly Applications. In R. Nazir (Ed.), *Nanotechnology Applications in Environmental Engineering* (pp. 308–343). Hershey, PA: IGI Global. doi:10.4018/978-1-5225-5745-6.ch013

Pesetskii, S. S., Bogdanovich, S. P., & Aderikha, V. N. (2019). Polymer/Clay Nanocomposites Produced by Dispersing Layered Silicates in Thermoplastic Melts. In N. Ramdani (Ed.), *Polymer Nanocomposites for Advanced Engineering and Military Applications* (pp. 66–94). Hershey, PA: IGI Global. doi:10.4018/978-1-5225-7838-3.ch003

Purohit, P., & Vagge, S. T. (2018). Oxidation Behavior of LTA/YSZ Intermixed Layer Coating. In A. Pakseresht (Ed.), *Production, Properties, and Applications of High Temperature Coatings* (pp. 107–130). Hershey, PA: IGI Global. doi:10.4018/978-1-5225-4194-3.ch005

Sadegh, H., Ali, G. A., Nia, H. J., & Mahmoodi, Z. (2019). Nanomaterial Surface Modifications for Enhancement of the Pollutant Adsorption From Wastewater: Adsorption of Nanomaterials. In R. Nazir (Ed.), *Nanotechnology Applications in Environmental Engineering* (pp. 143–170). Hershey, PA: IGI Global. doi:10.4018/978-1-5225-5745-6.ch007

Sadiku, E. R., Agboola, O., Mochane, M. J., Fasiku, V. O., Owonubi, S. J., Ibrahim, I. D., ... Ndamase, A. S. (2019). The Use of Polymer Nanocomposites in the Aerospace and the Military/Defence Industries. In N. Ramdani (Ed.), *Polymer Nanocomposites for Advanced Engineering and Military Applications* (pp. 316–349). Hershey, PA: IGI Global. doi:10.4018/978-1-5225-7838-3.ch011

Sadri, E., & Ashrafizadeh, F. (2018). High Temperature Nanocomposite Coatings by Plasma Spraying for Friction and Wear Applications. In A. Pakseresht (Ed.), *Production, Properties, and Applications of High Temperature Coatings* (pp. 216–245). Hershey, PA: IGI Global. doi:10.4018/978-1-5225-4194-3.ch009

Saeedi, F., Montazeri, A., Bahari, Y., Pishvaee, M., & Ranjbar, M. (2018). Synthesis and Characterization of Chitosan-Poly Vinyl Alcohol-Graphene Oxide Nanocomposites. *International Journal of Chemoinformatics and Chemical Engineering*, *7*(1), 1–12. doi:10.4018/IJCCE.2018010101

Shahien, M. (2018). Reactive Plasma Spray: A Method for Nitride Coatings Deposition in Thermal Spray. In A. Pakseresht (Ed.), *Production, Properties, and Applications of High Temperature Coatings* (pp. 299–332). Hershey, PA: IGI Global. doi:10.4018/978-1-5225-4194-3.ch012

Shahzadi, P. (2019). Properties of Nanomaterials and Environment. In R. Nazir (Ed.), *Nanotechnology Applications in Environmental Engineering* (pp. 28–43). Hershey, PA: IGI Global. doi:10.4018/978-1-5225-5745-6.ch002

Shamim, A., Ahmad, S., Khitab, A., Anwar, W., Khushnood, R. A., & Usman, M. (2018). Applications of Nano Technology in Civil Engineering: A Review. *International Journal of Strategic Engineering*, *1*(1), 48–64. doi:10.4018/IJoSE.2018010104

Shikuku, V. O., & Sylvain, T. (2019). Application of Geopolymer Composites in Wastewater Treatment: Trends, Opportunities, and Challenges. In N. Ramdani (Ed.), *Polymer Nanocomposites for Advanced Engineering and Military Applications* (pp. 131–149). Hershey, PA: IGI Global. doi:10.4018/978-1-5225-7838-3.ch005

Singh, J., Bhadane, M. S., Dubey, V., Dhole, S. D., Manam, J., & Dahiwale, S. S. (2018). Phosphors for Various Dosimetry Applications Derived by Different Synthesis Routes. In R. Tiwari, V. Dubey, & S. Dhoble (Eds.), *Emerging Synthesis Techniques for Luminescent Materials* (pp. 53–84). Hershey, PA: IGI Global. doi:10.4018/978-1-5225-5170-6.ch003

Som, S., Dutta, S., Kumar, V., & Swart, H. C. (2018). Swift Heavy Ion Synthesis and Modifications of Nanophosphors for Dosimetric Application: Effect of Swift Heavy Ion Irradiation. In R. Tiwari, V. Dubey, & S. Dhoble (Eds.), *Emerging Synthesis Techniques for Luminescent Materials* (pp. 1–25). Hershey, PA: IGI Global. doi:10.4018/978-1-5225-5170-6.ch001

Suresh, S. S., George, K., Mohanty, S., & Nayak, S. K. (2019). Architect of Polymer Nanocomposites for Aerospace Applications. In N. Ramdani (Ed.), *Nanotechnology in Aerospace and Structural Mechanics* (pp. 163–205). Hershey, PA: IGI Global. doi:10.4018/978-1-5225-7921-2.ch005

Tahir, L. (2019). Green Approaches to Environmental Sustainability. In R. Nazir (Ed.), *Nanotechnology Applications in Environmental Engineering* (pp. 81–101). Hershey, PA: IGI Global. doi:10.4018/978-1-5225-5745-6.ch004

Tripathi, S., Mishra, M. K., Jain, V. K., Tiwari, R., & Dubey, N. (2018). Thermo, Photo and Mechanoluminescence Studies of Eu3+ Doped Y4Al2O9 Phosphors: TL, PL, and ML Analysis of YAM: Eu3. In R. Tiwari, V. Dubey, & S. Dhoble (Eds.), *Emerging Synthesis Techniques for Luminescent Materials* (pp. 389–414). Hershey, PA: IGI Global. doi:10.4018/978-1-5225-5170-6.ch010

Ugemuge, N. S., Gayner, C., Natrajan, V., & Dhoble, S. J. (2018). Perspectives of Hydrothermal Synthesis of Fluorides for Luminescence Applications: Fluorides Phosphors for Luminescence. In R. Tiwari, V. Dubey, & S. Dhoble (Eds.), *Emerging Synthesis Techniques for Luminescent Materials* (pp. 277–303). Hershey, PA: IGI Global. doi:10.4018/978-1-5225-5170-6.ch008

V., S., R., M., S., K., & S., S. K. (2018). White Light Emitting Phosphors for Solid State Lighting. In R. Tiwari, V. Dubey, & S. Dhoble (Eds.), *Emerging Synthesis Techniques for Luminescent Materials* (pp. 150-250). Hershey, PA: IGI Global. doi:10.4018/978-1-5225-5170-6.ch006

Vaidyanathan, S., & Tagare, J. (2018). Phenanthroimidazole-Based Organic Fluorophores for Organic Light-Emitting Diodes: An Overview. In R. Tiwari, V. Dubey, & S. Dhoble (Eds.), *Emerging Synthesis Techniques for Luminescent Materials* (pp. 305–388). Hershey, PA: IGI Global. doi:10.4018/978-1-5225-5170-6.ch009

Van Miert, S., Creylman, J., & Verheyen, G. R. (2019). Mining a Nanoparticle Dataset, Compiled Within the MODENA-COST Action. *International Journal of Quantitative Structure-Property Relationships*, *4*(1), 1–17. doi:10.4018/IJQSPR.2019010101

Verma, R., Suri, N. M., & Kant, S. (2018). Slurry Sprayed Mullite Coatings and Their Corrosion Performances. In A. Pakseresht (Ed.), *Production, Properties, and Applications of High Temperature Coatings* (pp. 187–214). Hershey, PA: IGI Global. doi:10.4018/978-1-5225-4194-3.ch008

Xie, W. (2018). Thermal Spray Methods and Splat Formation. In A. Pakseresht (Ed.), *Production, Properties, and Applications of High Temperature Coatings* (pp. 1–24). Hershey, PA: IGI Global. doi:10.4018/978-1-5225-4194-3.ch001

Yaseen, T., & Yaseen, A. (2019). Wastewater Treatment in Removal of Heavy Metals: Nanotechnology Applications in Environmental Engineering. In R. Nazir (Ed.), *Nanotechnology Applications in Environmental Engineering* (pp. 220–240). Hershey, PA: IGI Global. doi:10.4018/978-1-5225-5745-6.ch010

Zhizhin, G. V. (2019). Higher Dimensions of Clusters of Intermetallic Compounds: Dimensions of Metallic Nanoclusters. *International Journal of Applied Nanotechnology Research*, *4*(1), 8–25. doi:10.4018/IJANR.2019010102

Zhu, S., & Wang, F. (2018). Nanocrystalline, Enamel and Composite Coatings for Superalloys. In A. Pakseresht (Ed.), *Production, Properties, and Applications of High Temperature Coatings* (pp. 160–186). Hershey, PA: IGI Global. doi:10.4018/978-1-5225-4194-3.ch007

About the Contributors

Amartya Chakrabarti is an assistant professor at Dominican University, USA. Dr. Chakrabarti received his PhD in chemistry from the Illinois Institute of Technology. He has (co)authored 13 research papers and 6 book chapters. He currently holds two patents, one of them is in the development of crystalline graphene. His current research focuses on developing novel methodologies to synthesize a variety of one-dimensional and two-dimensional nanomaterials for energy and space applications.

Nadeem Faisal is currently working as an Assistant Professor, in the Department of Mechanical Engineering, ITM University, Gwalior, India. He completed his Master's in Engineering from Birla Institute of Technology, Mesra, India. He has over 2 years of teaching and 1 year of Industrial experience. His areas of interests are Optimization, Materials Engineering, Product and Process Design, CAD/CAM, Rapid Prototyping and Digital Manufacturing. He has published 2 books, 14 Book Chapters, and research papers in 5 SCI & 4 SCOPUS Indexed international publication.

Siddhartha Kosti is working as an Assistant Professor in the Department of Mechanical Engineering, Rajkiya Engineering College, Banda, UP, India. He has completed his Masters from IIT Kanpur in Thermal and Fluid Science. He has 5 years of teaching and research experience. His expertise lies in the field of heat transfer, nanomaterials, composites, and CFD. He has published several refereed articles in reputed journals and conferences. He has also written a book chapter "SWOT Analysis of Literature" for the book 'Creating a Framework for Dissertation Preparation: Emerging Research and Opportunities' published by IGI Global. Recently he has received a research grant for a project related to solar PVT by NPIU-MHRD under collaborative research scheme.

Sukhila Krishnan completed his M.Tech degree in Polymer Science from Cochin University and his obtained PhD in Polymer Science and Technology at Laboratory for Advanced Research in Polymeric Materials (LARPM), Central Institute of Plastic Engineering and Technology, Bhubaneswar. Currently, she is working as an assistant professor at Sahrdaya College of Engineering and Technology, Kodakara, Thrissur, Kerala Her main research interests focus on biopolymers from natural resources.

Sudheer Kumar completed his M.Tech degree in Polymer Nanotechnology from Biju Patnaik University of Technology and his obtained PhD in Polymer Science and Technology from Utkal University Bhubaneswar, India at SARP-Laboratory for Advanced Research in Polymeric Materials (LARPM), Central Institute of Plastic Engineering and Technology, Bhubaneswar. His main research interests are the synthesis of epoxy resins from renewable resources, endcap oligomer and polymeric coatings.

Pradip Majumdar is Professor Emeritus and Emeritus chair at the Norther Illinois University. He is an expert in fluid mechanics, heat transfer, thermal energy systems, computational fluid dynamics (CFD) and heat transfer, and experimental methods, having over 34 years of teaching and research experience in simulation modeling and experimental analyses of wide range engineering applications including cooling of electronics and electrical devices, fuel cell and Li-ion Battery storage systems, engine combustion, transport phenomena in in biological systems, nanostructured thermal interface materials, and laser Material Processing and additive manufacturing processes. He has completed many funded research projects and published more than 120 journal and technical conference papers, authored three textbooks: Computational Methods for Heat and mass Transfer; Fuel Cells- Principles, Design and Analysis, and Design of Thermal Energy Systems (In Press). He has served as the Editor-in-Chief for the Journal of Transactions of Fluid Mechanics and Associate Editor of ASME Journal of Thermal Science and Engineering, and currently serving as the editor of the International Communications on Heat and Mass Transfer. He is professionally very active in organizing and chairing national and internal conferences, workshops in heat and mass transfer, computational fluid dynamics, micro-nanoscale heat transfer, electronics cooling and fuel cell energy systems. He has been making keynote and plenary speaker presentations at national/ international conferences and workshops.

Winfrida Nyairo is an Environmental-Analytical Chemist. Holds a PhD in nanomaterials science and serves as Lecturer at Maseno University, Kenya.

Ananthakumar Ramadoss is a Jr Scientist in the SARP-Laboratory for Advanced Research in Polymeric Materials, Central Institute of Plastics Engineering & Technology, India. He received his Ph.D. degree from Jeju National University, Republic of Korea in 2015 with major research on supercapacitors. He was a Alexander Von Humboldt Postdoctoral Fellow at the University of Konstanz, Germany and Postdoctoral fellow at the Korea Advanced Institute of Science and Technology (KAIST), Korea as well as a research fellow at the Central Electrochemical Research Institute (CECRI), India. He has published 1 patent and more than 45 peer-reviewed journal papers (h-index: 25) and 2 book chapters. His research interests mainly focused on the development of nanostructured materials and their applications in supercapacitors, nanogenerators, self-powered systems, corrosion protection and bio-implant applications.

Sushanta Kumar Samal obtained PhD in Polymer Science in 2009. He is now working as a scientist in SARP-Laboratory for Advanced Research in Polymeric Materials, Central Institute of Plastics Engineering and Technology, Bhubaneswar. His research interests include the development of new advanced materials, biopolymer, composite and nanocomposites, coating, foam, water purification membrane, etc.

Aruni Shajkumar is a research associate at the Laboratory for Advanced Research in Polymeric Materials, Central Institute of Plastic Engineering & Technology, Bhubaneswar, India. She received her PhD from Dresden Technical University, Germany in 2018 with a specialization in Hybrid Nanostructures. Post her PhD she was a guest researcher at the Leibniz Institute of Polymer Research, Dresden, Germany. She has 2 research papers in peer reviewed journals and 2 book chapters. Her research interests are mainly focused on developing multifunctional hybrid nanostructures and devising them multifunctional applications such as photocatalysis, smart materials, waste water treatment, biomedical and high-performance applications.

Victor Shikuku is a Physical, Materials and Environmental Scientist. Holds PhD in Physical Chemistry and serves as Lecturer at Kaimosi Friends University College, Kenya.

Margarita Tecpoyotl Torres received the Mathematician degree from the Autonomous University of Puebla (UAP), Mexico, in 1991. In this University, she was graduated as Electronic Engineer in 1993. She received the M.Sc. and Ph.D. degrees in Electronics from National Institute of Astrophysics, Optics and Electronics (INAOE), México, in 1997 and 1999, respectively. Dr. Tecpoyotl works, since

1999, at CIICAp of UAEM, Mexico, where she is currently titular professor. Her main research interest includes Electronics, MEMS and Microwave devices. She holds the status of National Researcher (SNI, level 1). She has 4 patents conceded by Mexican Institute of Intellectual Property, IMPI, and 2 under revision.

Rafael Vargas-Bernal received a bachelor's degree in Communications and Electronics Engineering from the University of Guanajuato in 1995, and the degrees of Master of Science and Doctorate in Sciences with Specialty in Electronics from the National Institute of Astrophysics, Optics and Electronics (INAOE) in 1997 and 2000, respectively. Since January 2002, he has been a professor-researcher at the Higher Technological Institute of Irapuato (ITESI) and, particularly since 2006, he has worked in the Department of Materials Engineering where he has established himself as a researcher. He has authored 15 articles in journals, 26 chapters in books and about 100 conference articles. He is a member of the National System of Researchers (SNI-Mexico). He regularly serves as a reviewer of scientific articles in RSC Advances and Royal Society Open Science as well as standards in Semiconductor Equipment and Materials International (SEMI). His research interests include two-dimensional materials, nanomaterials, space materials, composite materials, gas sensors and biosensors.

Danilo Vuono graduated in Chemical Engineering with a PhD in "Technology and environmental planning",. He worked at Nanopart, Leuven (Belgium) and at the University of Calabria, collaborating with important institutions such as, University of Manchester (UK), the KACST of Riyadh (Saudi Arabia) and the ITM-CNR (Italy). He has a large experience in the field of nanotechnology, with a strong background on synthesis of helical and supported nanotubes on various refractory materials and their application in the environmental field. Actually, he is Professor at Federiciana University, Italy, and he is CEO of Innovacarbon, an innovative startup working in environmental sector. His scientific production includes many papers on microporous materials and nanotechnologies published in international journals.

Index

www.ingramcontent.com/pod-product-compliance
Lightning Source LLC
LaVergne TN
LVHW061535290926
849162LV00004B/802